ISBN 978-0-265-99241-8
PIBN 10920299

MINES AND MINERALS

OF

WASHINGTON.

SECOND ANNUAL REPORT

OF

GEORGE A. BETHUNE,

STATE GEOLOGIST.

OLYMPIA, WASH.:
O. C. WHITE, . . . STATE PRINTER.
1892.

To Hon. Elisha P. Ferry, *Governor, and the members of the State Mining Bureau of Washington:*

Gentlemen — I have the honor herewith to submit my second annual report as state geologist of Washington.

With respect,

Geo. A. Bethune, *State Geologist.*

Tacoma, Wash., January, 1892.

PREFATORY.

In a prefatory way, it is my desire to address to you a short statement of facts, and also to digress here, from the subject in hand, that I may completely set forth circumstances, as I have been led to view them, which in the end have been productive, not alone of results injurious to the proper conduct of this office and incident official work therein, but have well nigh shut out and cut off all possibility of rendering useful and hence valuable the services of myself as your state geologist.

MINES AND MINERALS OF WASHINGTON.

PART FIRST.

SECOND ANNUAL REPORT OF GEORGE A. BETHUNE,
STATE GEOLOGIST.

PROPOSED GEOLOGICAL SURVEY.

At the session of the state legislature which convened in January, 1891, a bill was passed and ultimately accorded the approval of Hon. Charles E. Laughton, then acting chief executive of this state, whereby a certain sum of money, derivable from various sources of state revenue, was appropriated for the purpose of disbursing such sum in the preliminary work necessary to the commencement of a geological survey of the state, and for the purpose also of carrying ahead the actual work of such geological survey in so far as such appropriation would admit.

Acting upon the fact that this appropriation had passed both houses, and had been, therefore, recommended to the acting governor for his approval and approved by him, with the coöperation of the state mining bureau, I, as state geologist, completed the necessary preparatory work; and actual field operations were in progress, when the measure appropriating such sum for such geological survey was by our honorable supreme court declared illegal, therefore null and void. An unfortunate omission in the verbiage of the bill brought about this lamentable denouement to a plan certainly improvised for the benefit of our commonwealth, and one which, completed, no doubt would have been resultant in profit to our state, in even a far greater measure than the most sanguine could have anticipated.

CUMULATIVE RESULTS.

Cumulative results of the defeat of the plan of a geological survey of the state, I have here to say, were productive of great injury to this office; injury to its power for good, its reputation among those for whose especial benefit it was created, and to its general efficiency. In short, I feel at liberty here to say, that co-incidental with the defeat of a plan having for its object a geological survey of Washington, is presented the apparent fact that, for some time to come at least, the usefulness of a state geologist and the maintenance of a state mining bureau are at once at an end, and may in no way be made conducive to the public weal.

GEOLOGICAL PROCEDURE—1891.

Despite the facts I have hereinbefore recited, I have to report that, as during the previous year, I have given my undivided attention to the demands of the state from a mineralogical standpoint, in so far as I have been, alone and unaided, able to do. Despite hinderances innumerable, annoyances indescribable, and laboring under a stigma attached to both the mining bureau and this office by reason of events of a character political and in no way attributable to the conduct at least of this office, I have pursued my work without cessation, and in order that the affairs and operations of this office may be more fully illustrated to you, append this, my second annual report.

GROWING INTEREST.

I have to report a wide interest apparent among our populace and that abroad in our mineral resources which, during the past twelve months, have been accorded a very fair measure of development. The infusion of a much larger amount of capital in the project of developing and rendering productive these resources has, as a result, not alone rendered such development possible on a much grander scale than formerly, but has invited a renewal of interest on the part of everybody in our mines and minerals—an interest, I am happy to state, which now gives promise of being indefinitely retained. Several new and important mineral discoveries in widely separated and hitherto practically unprospected regions have been made, and all give promise of a speedy attainment to a position of wealth and productiveness, and a place among the first of our mineral bearing areas. Especial attention has been paid by me to the coal measures of our state, and I invite your attention to the following chapter regarding them.

MINES AND MINERALS OF WASHINGTON.

PART SECOND.

SECOND ANNUAL REPORT OF GEORGE A. BETHUNE,
STATE GEOLOGIST.

WASHINGTON COALS.

In considering the coal measures of Washington, I shall, in a prefatory way, briefly outline as it is presented to me the geological aspect of the state, with a view of at once more completely illustrating and elucidating the arguments and statements herein made, and of natural events which finally led to the formation by natural process of the coals which in this state abound, and which are destined to enact a leading part in its general upbuilding and advancement.

Washington State appears to the geologist, as a result of inspection conducted from a topographical standpoint, a panorama of low land, plateaus of greater or less altitude, and mountains of greater or less height. An extended account of the topographical aspect I deem here not necessary, in that it is well known, both as the result of private and governmental investigation, often times repeated; described by eminent authority, and known to the general public, both within and without the state.

Taking first the mountainous country constituting a considerable portion of the area of the state, it is agreed generally by those who have studied the formation, that the rocks composing it may be either of the lower cretaceous, palæozoic or archaen series, or may be classed as the metamorphosed strata of the upper juro trias.

The fact that the core of the various mountain ranges is mainly composed of originally stratified rock, metamorphosed by heat, lends color to the correctness of a theory that the rocks composing them are but the metamorphosed strata of the upper juro trias.

Undoubtedly of plutonic origin are rocks forming branches thrown out in different directions from the main ranges, which rocks have never undergone the process of stratification. This condition is certainly apparent along the summit of the Cascade range of mountains, being most plainly discernible, I take it, in the vicinage of Snoqualmie Pass, in the northeastern corner of King county, in Western Washington, and also along some of the minor vertebræ, owing existence to the upheaval of the Cascade or mother range.

In several localities along this range where greater power has elevated to a higher altitudinal plane, portions of the range, Mts. Baker, Denny, Rainier, Hood, etc., the period of formation of which is still a question for study, is found syenitic granite in enormous deposits; with this syenitic granite are found associated chloritic rocks carrying magnetic iron ores, crystalline lime stone beds, both of fine and of course texture, extremely hard quartzites of fine grain, and various porphyries and serpentoids.

Veins containing metals of both the precious and base varieties are found to cross these quartzites, porphyries and serpentoids at various angles.

In the more elevated ranges lying north of the Columbia river, these plutonic and metamorphic rocks are to be found. It has been demonstrated that on both the eastern and the western sides of the Cascade mountains are sand stones and cretaceous conglomerates, and as far as development has progressed, shales, carrying coals in their upper beds.

An almost total denudation characterizes the deposits found on the eastern side of the Cascade mountains, where the deposits are found to be considerably broken by arroyos, or mammoth ravines and gulches. As a result, areas of comparatively diminutive size, as regards the cretaceous strata, are to be found on the Yakima and the Peschastin rivers, and that great spur of the Cascade mountains known as the Peschastin Ridge, or divide, which lies between the streams named.

Because of the presence of no grooving of the region forming the west side of the Cascade mountains, these areas are found to be much larger in extent.

While the coal deposits found west of the Cascade mountains are popularly looked upon as related to that period — the finalé, so to speak — of the cretaceous era, which is termed the Laramie period, constituting a progressive era between the cretaceous and the tertiary regimes, I cannot authoritatively here chronicle the fact of the demonstration beyond peradventure of the correctness of this widely accèpted conclusion. Such practical demonstration has, however, been made, as regards the lower order of lignites found in that main vertebrae called the Rocky mountain range, and the semi-lignites or off-shoots of the lignite species, constituting the highest altitudinal beds of coal in the state. The classifying of the underlying beds of coal here found, and the placing of them in relationship also with the Laramie period, must, however, remain still a matter of question, meet for future and extended and careful investigation.

The beginning of the coal bearing rocks of Western Washington may be traced from the western base of the Cascade mountains and on the outlying spurs of that range. As a generality these outlying spurs in size, height and longitudinal and latitudinal tendency are decidedly irregular. Their altitudinal proportions may be safely approximated at from 800 to 2,200 feet. They are found in groups, usually very closely huddled together, separated by the denuded areas lying between the main range and Puget Sound and the ocean on the west, from the boundary outlining the British possessions on the north to near unto the Columbia river on the south. The largest coal field in the state lies within these boundaries. In the southern portion of this field the majority of the measures appear to me to belong to the tertiary era, as those also evidently do which are to be found along the Chehalis, Cowlitz, Nesqually and Skookum Chuck rivers.

Conspicuous features of the Cascade mountains are its lofty volcanic peaks which, it is practically settled, must have been active volcanoes during the tertiary epoch, thereby not only upbuilding their summits to altitudes as high as 14,400 feet, but immersing the areas contiguous to them in a bath of moulten lava of great size and depth. Mts. Baker, Rainier, St. Helens, Hood and Adams seem to me to have been the principal theaters of these volcanic dramas.

Most noticeable results of this active volcanic era are to be found in the great arroyo carved out of lava by the Columbia river in its course to the sea through the Cascade mountains. There the volcanic matter is found to comprise chrysolite and diabase which have cooled into basalts and chrystallizations, which latter, so wrought by the climatic action of ages, are there to be found in all sorts of fantastic forms, many of them being of rarest beauty.

Undoubtedly successive inundations of lava have had much to do with the upbuilding of these deposits of basalts and chrysolites, the joints of which are in many places plainly discernible. For hundreds of miles up the Columbia river from the point of passage through the Cascade mountains will be found these deposits, they averaging from 650 to 1,800 feet in thickness. Eruptions and incident inundations of lava on both the eastern and western sides of the Cascade mountains have occurred, with greater or less force, at a comparatively recent period.

What is termed the quaternary, more popularly designated the drift, period, has left us clearly definable proofs of its presence. These evidences are mostly abundant in Western Washington, there being comparatively few traces of the quaternary period to be found in the eastern portion of the state; they being mainly confined, in fact, to a small deposit of quartz gravel forming the irregular shaped flats bordering the Spokane river.

Drift material, plainly denoting in its appearance the action of both ice and water, is to be fonnd in a number of places in the rolling country lying to the north and east of the waters of Puget Sound; rotund boulders of granite of greater or less size being found on nearly all of the more elevated areas bounding the Sound shore line. The low lands in the territory described are found remarkably free of these traces.

Restricting further remarks of a geologic character to the limit of their application to our coal measures, I will begin with the statement that in so far as the measures of the great Puget Sound basin are concerned, their thickness approximates 13,500 feet. Of course, a sufficient measure of development not having yet been accorded these measures, nothing nearly likened unto an accurate statement in this regard is possible. Taking into consideration the fact, as hereinbefore stated, that the deposits alluded to grow lesser in depth from west to east, the approximate thickness of the measures stated, *i. e.*, 13,500 feet, may not be expected in every section of the area. Numerous fractures and differences as regards the strike of the measures also tend to preclude an accurate measurement.

I consider that the number of separate, distinct and workable seams of coal in these measures, as far as developed, and which are not less than three and one-half feet in width, does not exceed twenty-three.

A cause for regret is the stinted nomenclature accorded the coals of the Puget Sound basin, also the unjust (if not improper) classification assigned some of these coals by very reason of the limited boundaries of this nomenclature. It is only recently that the coals of this basin, plainly traceable to a relationship with the Laramie and cretaceous eras, have been classified into the lignite, semi-bituminous, bituminous and anthracitic divisions. These are in this basin found to grade into one another so imperceptibly as to preclude all possibility of their acute classification; and, as an example of the total unfitness of our nomenclature to serve the purpose intended, let me cite the fact that the discovery remains to be made of a species of lignite in this basin as low in grade as the universally accepted lignite coal. It will be seen that a more diversified nomenclature should be made for our coals, that a good coal may not be made to suffer from classification under an inferior title. Analytical demonstration shows that in the main Washington lignite coals hold a larger percentage of oxygen than do the average coals from elsewhere, which latter are boldly placed in the bituminous division of a, so to speak, foreign and more generous nomenclature. These should be accorded a more advanced classification, therefore; one that would place them on a parity with at least what are popularly accepted bituminous coals.

Lignites, semi-bituminous and bituminous coals comprise the species found in the measures of the Puget Sound basin. I have yet to hear of any anthracite coals being discovered.

The lignite measures extend from the county of Whatcom on the north, through King, Pierce, Thurston, Cowlitz and Chehalis counties to the southward. Other strata of lignites (evidently off-shoots) exist on the Skookum Chuck river, in what is called the Hannaford Valley region. The workable seams of lignite so far uncovered in the great basin do not exceed eighteen in number.

The bituminous and semi-bituminous measures of the basin are to be found below and to the eastward of the lignites. These bituminous measures may be said to extend from Lake Whatcom (Whatcom county) clear into Lewis county on the south.

From a commercial standpoint, the lignite coals of the basin are of most importance. The area in which these coals are found is also larger than those combined in which have been found existent the other species. The extent of the lignite area has been estimated at from 650,000 to 675,000 acres. The workable coal seams actively operated and undergoing process of development in the lignate area number twenty-three, with scenes of operations located from Whatcom county through to the Cowlitz river.

Principal among those seams which have been made productive

are those at New Castle, on the line of the New Castle branch of the Columbia & Puget Sound Railway, in eastern King county, known as the New Castle mine, the property of the Oregon Improvement Company. Here the mining of six distinct seams has been in progress for several years. The trend at the seams is south-southeast and north-northwest, with a dip of 45 degrees to the north.

At Gilman, on the line of the Seattle, Lake Shore & Eastern Railway, and distant twelve miles from New Castle, and also at Coal Creek, distant three and one-half miles from New Castle, are found workable seams of lignite coal, which have been mined with profit for some years. At Gilman, what is termed the Gilman mines, a seam of clean lignite coal eleven feet in width is being mined. This is the property of the Seattle Coal & Iron Company. At Coal Creek operations were successfully conducted by the Oregon Improvement Company for three years on a seam of clean coal eight feet nine inches in width, but for a reason unknown to me were discontinued six months ago.

Other seams from which coal is being regularly mined are found at Black Diamond, forty-three miles southeast of New Castle; Franklin, three and one-half miles southeast of Black Diamond; New Green River Colliery, three and one-half miles southeast of Franklin; Lizard Mountain Colliery, and at Durham, three and one-half miles northeast of Lizard Mountain; at Kangley, one and one-half miles northeast of Durham; at Alta, one and one-half miles northeast of Kangley; at Ruffner, near the headwaters of Raging river, and at Cedar Mountain, all in King county. Seams not producing at present, but developed, and seams in process of development, are to be found near the town of Renton (Renton C. M. Co.'s mine), at Bucoda, at Cherry Hill, in Lewis and Chehalis counties, and along the Cowlitz river and its tributaries. A pronounced similarity in trend and dip will be found in all the seams I have mentioned. The measures found, however, in the southeastern portion of the basin are plainly of more recent origin than those found in its northern area.

On the straits of Juan de Fuca, in northern Clallam county, is found another series of lignite measures, extending from a point near Clallam Bay eastward to Pillar Point on Puget Sound. This series, so far as developed, comprises three workable and five "dirty" seams of coal. One is undergoing process of development. The coals resemble the New Castle product, but, I am of opinion, are not of the same age.

In Eastern Washington are found more lignite measures. The workable seams thus far exposed in this region number three. The measures begin at the head waters of the Methow river, in Okanogan county, trending in a southwesterly direction. Aside from a small measure of prospect work accorded these seams, they have undergone no development of moment.

In the Wenatchee river region, Okanogan and Kittitas counties, are found lignite measures, but nothing has been done in the way of their development.

BITUMINOUS COALS.

The styling of a species of coal bituminous is popular, if in so doing a misnomer is created, for it is the fact that there is no bitumen in coal. Possibly, because of the fact that when heated certain coals liberate a portion of the volatile matter contained in them in the shape of an opaque, oily liquid resembling bitumen, this title has been given them.

A "bituminous" coal is given classification as such, however, because of its percentage of volatile matter, which usually exceeds 20 per cent. The specific gravity of the "bituminous" species ranges from 1.20 to 1.30. In semi-"bitumious" coals the specific gravity will range from 1.30 to 1.45, and the volatile matter will be found to range from 10 to 20 per cent. While, as I have stated, in our nomenclature the classification of our coals into distinctive divisions is rendered practically impossible, bituminous coals are more widely divided elsewhere; and, as these divisions are locally applicable, I shall note them.

Species of the bituminous variety are known as coking (caking) coals, cherry coals, split or black coals (dry burning or open coals), and cannel (candle) coals.

Seams of "bituminous" coal adapted to the manufacture of coke are found in Western Washington in Skagit, King, Pierce and Whatcom counties, and in Eastern Washington in Kittitas county, along the head waters of the Yakima river. These coals, in process of coking, fuse their fragments together in an adhesive mass, and as the gases are liberated by the heat, this mass is puffed up into a hard and highly cellular substance consisting of fixed carbon and the mineral matters in the coal.

The split or black coals, while "bituminous" in character, are diametrically opposite to the coking coals, in that their transformation into coke cannot be made to occur.

These coals have thus far been found only in Kittitas county, though undoubtedly further investigation will disclose their presence elsewhere in the state. In the manufacture, or rather in the work of reducing our iron ores to the metallic state, these coals will be found well adapted.

I have seen excellent samples of cannel coal from the Skokomish river, in Mason county. This coal is best adapted to the manufacture of gas because of the high percentage of its volatile gases.

The "bituminous" measures of the Puget Sound basin are found to be bounded to the westward by the lignite belts, the former being between the latter and the Cascade mountains. Characteristics of structure and chemical composition are found to be radically different from those noticeable in the lignite measures.

The "bituminous measures" we find tilted at angles much more elevated than are those noted in the lignite measures, the dip of the bituminous measures varying generally from 57 degrees to 63 degrees. In the work of arriving at a conclusion as to the primary state of our "bituminous" measures, and also at a determination of the number of workable seams in them contained, the geologist will find himself at a loss to compute accurately in these regards, because of the warped character of these veins. The system of mining the bituminous coals of the Puget Sound basin is also of such a character as to, if not wholly preclude, at least let and hinder to a degree scientific research into the origin and numerical strength of the seams through this channel, thus again precluding the furnishing of accurate information on these points.

I shall here now first consider seams at present productive, from which are being mined coals of a coking character, commencing at the seat of operations furthest toward the north in the great basin.

THE FAIRHAVEN MINES.

These mines are located twenty miles southeast of the city of Fairhaven, in Whatcom county, northern Western Washington. They are the property of the Skagit Coal & Transportation Company. The seams are on the "Whatcom measures," number eight; four of which are workable. The width of these are, respectively, nine feet, twelve feet and nine inches, eighteen feet and six inches, twenty-one feet and one inch, and twenty-three feet. All four have been cut by a tunnel 1,200 feet in length. The coal is eminently well adapted to the manufacture of a coke of fine quality. (Note analysis.) The company is erecting coking ovens. The mine is easily accessible by rail.

CONNOR-CUMBERLAND MINES.

The seams here found may be said to be comparatively new discoveries. The property is that of J. J. Connor and the Cumberland Coal Company. The seams number eleven, of which six have thus far been demonstrated workable. They are located on the south side of Skagit river, near the town of Hamilton, in Skagit county, and may be reached by all rail route from either of the principal Sound railway termini. Development work thus far completed comprises a level run in and tapping six seams, the last with an overhead lift of 370 feet. At the croppings these seams measure, respectively, in clean coal, five feet, four feet, five feet and six inches, three feet and six inches, three feet, and three feet and four inches. Coal from the Connor seams, the sextet alluded to, cokes splendidly, and may be termed friable.

THE BENNETT MINE

Is also located on the Whatcom measures. The scene of operations here is distant seven miles from Hamilton, and may also be reached by an all rail route. The Bennett is operating three seams of clean coal, three and one-half feet, three and one-half feet, and four feet and seven inches, respectively. The coals mined are identical with those found in the Connor-Cumberland seams, and their trend and dip correspond.

THE PAUL MOHR SEAMS

Is the accepted title to what is really the property of the Snoqualmie Coal & Coking Company.

No development of the measures extending south from the Bennett mine is to be noted until these mines are reached. They are situated on the Snoqualmie river, at a point distant from Snoqualmie Falls about seven miles, and from Seattle about seventy miles by rail. These seams belong to what we term the Snoqualmie division of the Puget Sound basin's bituminous coal measures. Here are found in process of development six seams of workable coal, one only of the coking variety. The trend of the seams is north and south, with a dip toward the west. The width of the coking seam is seven feet and six inches, and along its trend a tunnel has been driven a distance of 170 feet. The coal mined is a black, bright, laminated coal, some of which is easily disintegrated. On the same measures are numerous prospects, some on the Mohr coking seam described, among which latter are the properties of Messrs. Allison, Lee, Edwards, McAnders, and Nasson.

THE WILKESON GROUP

Is the popular designation given nine seams of coal being mined in the vicinity of the southern extremity (in so far as the Puget Sound basin is concerned) of the bituminous measures. This group lies in the southeastern portion of Pierce county; its members are accessible by rail, both from Portland, Oregon, and the principal points on Puget Sound.

Characteristics of this division of these measures noted here are the high angles at which they stand, and the lack of uniformity of dip. Taken topographically, the area in which is located this group may be described as anti-clinal, and in shape resembling a dial with its surface very highly elevated; its flanks being compressed and greatly grooved, with the major axis trending almost due north and south. As regards the situation of the members of this group, with reference to this major axis, I will state that the Wilkeson mine, known as Wilkeson No. 1, and the White mine, known as Wilkeson No. 2, and the South Prairie or Burnett mine, are located directly upon it; while the Carbonado mines, among the most extensive in the basin, are found to the southward of this axis. The anti-clinal character of the topography of this area is found to extend to the Nesqually river, where other measures, apparently forming a separate series, are to be found. The coals contained in them closely resemble those of the Wilkeson group of the major axis.

Undoubtedly due to seismic action, this coal area is found to contain a quality of "bituminous" coal eminently well adapted to the manufacture of coke of superior quality. Of the mines mentioned those producing coking coals are the two Wilkeson properties. The seam in the Wilkeson mine (No. 1), now being mined, is five and one half feet in width. The coal extracted is a light, bituminous coal, soft and friable. The seam in the White mine (Wilkeson No. 2) is six feet and four inches in width. The general appearance, quality, etc., of the output noticed in Wilkeson No. 1 are here reproduced.

To the north and east of the coking coal producers named lie several very promising prospects, all evidently on the mother measure. Among these I may mention those of the Lackawanna Coal Co., the Maguire seam, the Thompson-Davis prospect, the Waters prospects, and many others. The showings in all are of good character.

GAS COALS.

Many of the "bituminous" coals of the Puget Sound basin have been found as valuable for the manufacture of gas as for steam and domestic purposes. While they may be said to serve the former purpose best, coals found in the southeastern portion of the basin rank superior; still, some coals equally well adapted to gas manufacture have been mined these several years in other portions of it. One of the Paul Mohr seams alluded to contains an excellent gas coal. In three of the workable seams at Ruffner, on Raging river, are found good gas coals. These seams are only just being made productive.

At the South Prairie and Carbonado mines are being worked two seams — six feet four inches and seven feet two inches, respectively; the former of which produces an excellent gas and the latter a fine quality of steam coal. There can be no doubt that even some of the lignite seams contain coals suitable for the manufacture of gas. The product of the Alta, Kangley, Green River, Gregg, Franklin, Black Diamond, Cedar Mountain and New Castle seams will be found of a character suitable for the purpose.

The Blue Canon mine, just being made productive, and which is situated on the Whatcom measures, should produce a first-class gas coal. The trend and dip of its three workable seams is similar to those noted in the measures. At a depth of seventy-five feet a seam eighteen feet in width of clean coal is exposed. At a depth of 215 feet the seam is found to be twenty-three feet seven inches in width, clean coal. The Occident and Blue Canon companies are operating the same seams, the product closely resembling that of the Black Diamond mine, eighty-five miles to the southeast.

STEAM AND DOMESTIC COALS.

The area of the actual coal producing territory of Washington has been variously estimated at from 12,000 to 22,000 acres. It is my opinion that about 18,000 acres would closely approximate its extent. The extent of territory, however, underlaid with coal measures cannot be even approximately estimated, for reasons to which reference has already been made. It has been set down at 1,000,000 acres, and at 1,100,000 acres, and at 1,250,000 acres. I accept these estimates as simply the guess-work of their manufacturers.

As I have stated, coals of all species and classified members of each are to be found in our coal fields, anthracites and blacksmith coals excepted.

Coals suitable for steam and domestic purposes are found in nearly every workable seam in the Puget Sound basin, and in hundreds of new prospects. Of the operating mines I have mentioned, all are producers of steam and domestic coals of greater or less excellence. In the southern and western portions of the basin, producers of steam and domestic coals number every mine in operation, four now idle, and dozens of prospects awaiting the advent of capital to aid in their development. Both the Wilkeson mines, the Bucoda, Larabee, Cherry Hill, Carbonado, South Prairie, Lizard Mountain, Durham (idle), Kangley, Alta, McGuire, New Green River, Ruffner, Lackawanna, Franklin, Black Diamond, Cedar Mountain (idle), Renton (idle), New Castle and Gilman and Paul Mohr seams are all carriers of coals adapted, to a greater or less degree, to the manufacture of steam; and they are all also well known to be valuable for domestic purposes.

IN EASTERN WASHINGTON.

Eastern Washington—"the grand plateau of the Columbia"—may be described, topographically, as a rectangular table-land, the Cascade mountains, main axis of the state, constituting its west side—irregular spurs jutting out from its range along the British Columbia boundary line constituting its northern boundary, with Idaho as the eastern boundary. Principally in Oregon, but partly in this half of Washington, are the Blue mountains, which may be said to represent the south boundary of the area.

A series of coal measures exists in that part of the area described, comprising the westernmost portion of Kittitas county; and the largest mine in the Pacific Northwest, continually producing a superior quality of gas and domestic coal, is to be found in operation on about the central division of these measures. This is known as the Roslyn mine, the property of the Northern Pacific Railroad Company, and its chief source of supply west of Red Lodge, Montana.

The field in which this mine—or mines, for there are four separate and distinct workable seams being mined—is situated lies on the east flank of the Cascade mountains, along the Yakima river and its principal tributaries, the Cle-Elum and Teanaway streams. The general dip of the area will not exceed from ten to eighteen degrees, and the measures comprise four seams, all workable, and in width five feet seven inches; five feet seven inches; seven feet five inches, and six feet eight inches, respectively. The thickness of the coal bearing rocks is estimated at 1,250 feet. Undoubtedly these measures form the lower series.

While not a coking coal, the Roslyn product ranks *the* steam coal thus far mined in this state. It has been used as such with unqualified success by the Northern Pacific Railway Company for several years.

The Roslyn seam No. 1 carries five feet and seven inches of "bituminous" coals, which are jet black, hard, and free burning. These coals have been introduced with success the past year in Oregon, Idaho and Montana, and as far to the eastward as Minnesota and the two Dakotas.

The measures of Roslyn are those lying along the Cle-Elum and

Teanaway rivers. Those lying along the Wenatchee undoubtedly belong to the same series, but the seams have as yet undergone but little development. The coals resemble closely the Roslyn product. One seam has been exposed a considerable distance, which shows eight feet and four inches of clean coal. Coal-bearing rocks are traceable for a distance of forty miles up the river, their thickness approximating 91,080 feet.

MINES IN OPERATION.

Below will be found data concerning mines in actual operation in the Puget Sound basin and in Eastern Washington.

MINE.	SUPERINTENDENT.	COUNTY.	Number men	PER DIEM PAY.		
				Top men.	*Co. men.*	*Good pickmen.*
Roslyn	S. Reynolds	Kittitas	385	$2 50	$3 00	$6 00 to $6 50
Carbonado	T. D. Davies	Pierce	215	2 50	3 00	4 00 to 5 00
Wilkeson No. 1	A. C. Kelly	Pierce	170	2 50	3 00	4 00 to 5 00
Wilkeson No. 2	J. C. White	Pierce	90	1 75 to $2 50	3 00	4 50 to 5 50
South Prairie	C. H. Burnett	Pierce	85	2 00 to 2 50	2 50	4 00
Kangley	John Kangley	King	63	1 75 to 2 00	2 50	4 00
Alta	J. C. Anderson	King	24	2 00 to 2 50	2 50	5 00 to 6 00
New Green River	A. P. Williams	King	40	2 00 to 2 50	2 50	4 00 to 6 50
Franklin	T. B. Corey	King	210	1 50 to 1 75	2 00	4 00 to 6 50
Black Diamond	H. Morgans	King	185	2 00 to 2 25	2 50	4 00 to 6 50
New Castle	L. E. Wright	King	225	1 50 to 1 75	2 00	3 00 to 5 50
Gilman	Col. Parker	King	120	1 50 to 1 75	2 00	3 00
Blue Canon	J. Downs	Whatcom	35	2 00 to 2 50	2 50	3 00
Connor-Cumberland	J. J. Connor	Skagit	5	2 00 to 2 50	2 50	3 00 } Day work.
Bennett	E. P. Jennings	Skagit	15	2 00 to 2 50	2 50	3 00
Lizard Mountain	J. McGough	King	15	2 75	3 00	3 50

LIGNITE COALS.

I append results of analyses of lignite coals of the Puget Sound basin.

Name of Mine.	*Moisture* ...	*Vol. Com. Matter*....	*Fixed Carbon* ...	*Ash*	*Sulphur*	*Total*........
Gilman	4.27	39.02	49.48	7.00	.23	100.00
New Castle..............................	2.12	43.90	49.70	4.15	.18	100.00
Black Diamond.......................	3.11	45.10	47.20	4.57	.02	100.00
Franklin	2.98	43.05	50.00	3.95	.02	100.00
Kangley..................................	5.00	39.43	47.45	8.02	.10	100.00
Ruffner..................................	4.82	37.12	46.02	12.04	.00	100.00
Durham..................................	3.00	37.40	49.12	10.38	.10	100.00
Cedar Mountain.....................	3.24	43.52	48.39	4.85	.00	100.00
Pittsburgh	2.50	48.71	45.37	3.42	.00	100.00
Renton	3.24	39.52	51.40	5.76	.08	100.00
Bucoda..................................	8.10	42.95	35.56	13.20	.19	100.00
Cherry Hill............................	10.00	43.00	33.00	14.00	.00	100.00
Larabee	3.80	43.70	48.30	3.20	1.00	100.00
Clallam County.....................	9.98	40.53	40.07	9.16	.26	100.00
Methow	2.50	43.71	49.27	4.26	.26	100.00
Calispel	2.39	41.18	42.92	13.21	.30	100.00
Kallin prospect......................	3.10	43.00	51.00	2.90	.00	100.00
Thompson.............................	3.95	38.44	48.60	8.95	.06	100.00
Davis-Sjodin.........................	4.00	38.39	48.13	9:38	.10	100.00
Lee-Everson..........................	3.82	48.00	48.14	.10	.04	100.00

BITUMINOUS COALS.

Appended will be found a table of analyses of bituminous coals of the Puget Sound basin and Eastern Washington:

Mine.	*County.*	*Moisture*	*Vol. Com. Matter*	*Fixed Carbon*	*Ash*	*Sulphur*	*Total*
Cumberland	Skagit	.52	18.68	70.24	9.61	.95	100.00
Bennett	Skagit	.98	19.20	69.35	10.34	.13	100.00
Blue Canon	Whatcom	3.00	29.47	60.12	7.35	.06	100.00
Mohr Seams	King	3.90	34.28	60.53	1.10	.19	100.00
Wilkeson No. 1	Pierce	1.33	25.88	66.75	6.04	trace	100.00
Wilkeson No. 2	Pierce	2.50	26.30	64.90	6.25	.05	100.00
South Prairie	Pierce	3.01	32.81	60.02	4.06	.10	100.00
Carbonado	Pierce	1.80	42.27	53.11	2.82	.00	100.00
Hogan Bros	Pierce	2.00	40.39	53.50	4.11	.00	100.00
Lizard Mountain	King	2.10	25.57	70.15	2.18	.00	100.00
Lackawanna	King	3.51	30.00	60.10	6.39	.00	100.00
McGuire	King	2.30	29.90	61.00	6.80	.00	100.00
Lee & McAndrews	King	3.00	34.00	54.09	8.90	.01	100.00
Roslyn	Kittitas	3.15	35.70	53.00	8.02	.13	100.00
Frank Cook Seam	Kittitas	4.12	43.00	45.58	7.30	.00	100.00
Skagit Coal & Transportation Co.'s Mine	Whatcom	.94	29.00	60.00	10.00	.06	100.00

COKE ANALYSES.

Note below analyses of coking coals of the Puget Sound basin, and result of analyses of coke from the newly discovered Frank Cook seam, the only coal adapted to the manufacture of coke thus far found in Eastern Washington.

Mine.	*County.*	*Com. Carb.*	*Ash.......*	*Sulphur....*	*Phos-phorus...*	*Total.......*
Connor..	Skagit..................	87.21	11.87	.90	.02	100.00
Mohr seam..................................	King........................	90.18	9.37	.45	trace	100.00
Tacoma Coal & Coke Co........	Pierce..................	91.37	8.53	.10	.00	100.00
Wilkeson..................................	Pierce..................	89.90	10.00	.10	.00	100.00
Alta..	King......................	87.13	12.50	.37	.00	100.00
Lizard Mountain....................	King.......	90.18	9.37	.45	.00	100.00
Cook seam................	Kittitas................	79.90	19.63	.47	.00	100.00
McGuire seam.	King......................	78.30	21.20	.50	.00	100.00
Ruffner....................................	King...	90.40	9.48	.12	.00	100.00
Lizard Mountain......................	King....	90.18	9.37	.45	.00	100.00
Skagit Coal & Transportation Co.'s mines................	Whatcom...........	87.80	11.12	.06	.02	100.00

COMPARATIVE ANALYSES.

As a matter of reference and an item of interest, I append this table of analyses of lignite and bituminous coals and cokes from other regions:

LIGNITE COALS.

Location.	Moisture	Vol. Com. Matter	Fixed Carbon	Sulphur	Ash	Total
Bovey, England	6.20	22.86	56.31	2.36	12.27	100.00
Bel Monte, Colo	20.00	29.30	38.70	1.13	10.87	100.00
Grand River, Dak	13.11	38.90	34.70	.29	13.00	100.00
Bozeman, Mont	12.00	39.02	35.91	.15	12.92	100.00
Green River, Wyo	8.03	34.01	47.90	.10	9.96	100.00

BITUMINOUS COALS.

Location.	Moisture	Vol. Com. Matter	Fixed Carbon	Sulphur	Ash	Total
Maryland	1.25	15.80	73.01	.00	9.94	100.00
Pennsylvania	.82	17.01	68.82	.00	13.35	100.00
Virginia	1.64	36.63	50.99	.00	10.74	100.00
Ohio	7.20	30.10	57.55	.00	5.15	100.00
Indiana	2.20	33.99	58.44	.00	5.37	100.00
New Castle, Eng	6.06	9.91	80.03	.78	3.22	100.00
Ash, Eng	7.13	23.79	62.57	.40	6.11	100.00

COKE.

I append, as a comparison, analyses of coke from without the state.

Location.	*Com. Carbon.....*	*Ashes.......*	*Sulphur.....*	*Water.......*	*Phos-phorus....*	*Total.. .....*
Pennsylvania..........................	89.94	9.11	.82	.03	.10	100.00
Colorado	88.60	9.12	1.00	1.28	.00	100.00
Missouri.	90.00	8.70	.98	.32	.00	100.00
Indiana	91.60	7.66	.56	.18	.00	100.00
Durham, England	95.20	4.40	.30	.10	.00	100.00

Leaving a comparison of the foreign with our domestic analyses of lignite and "bituminous" coals to the inspection of the reader, I also conclude, without comment, this necessarily brief resume of Washington's coal deposits, which I believe rank among her leading resources.

MINES AND MINERALS OF WASHINGTON.

PART THIRD.

SECOND ANNUAL REPORT OF GEORGE A. BETHUNE,
STATE GEOLOGIST.

WASHINGTON IRON ORES.

We know of iron deposits being in existence in no less than twenty-two political subdivisions of Washington, *i. e.*, in King, Pierce, Thurston, Lewis, Cowlitz, Chehalis, Mason, Pacific, Kitsap, Jefferson, Clallam, Snohomish, Skagit and Whatcom counties, in the great Puget Sound basin; and in Kittitas, Okanogan, Stevens, Lincoln, Clarke, Klickitat, Whitman and Yakima counties in Eastern Washington.

It may be said of our iron ore deposits, therefore, that they are co-extensive with our coal measures; commensurate as regards area in which they are to be found with our gold and silver territory, unrivalled as to this relationship by any other series of deposits known to our times.

As varied in character as extensive in area are these deposits. Washington may boast possession of a great variety of iron ores. To my mind one of the most wealthy of our resources (and their name is legion) is to be found in our practically limitless deposits of these ores.

Among the rank and file of the base metals, from the very fact of its absolutely indispensable relation to innumerable forms of human industry, iron must be accorded the leading place. Iron is the metal of which it may be said that the progressive civilization of the day, with its countless and yet rapidly increasing needs, cannot abide without.

To the method of occurrence of the ore carrying this metal I shall briefly refer. Iron ores of these United States may be said mostly to be found in bedded deposits, sometimes scattered through these beds like the carbonate ores; sometimes occupying the major portion of the strata which, as is the case with them in Washington, are sometimes of great size. The ore beds have much the appearance of veins when found in association with strata, thrown into acutely inclined attitudes, and which have been altered to a very considerable degree. This is the case with the iron deposits of this state. That the strata found here were accumulated as part of the regular series of events which also had to do with the formation of these veins, and that, despite their appearance, they are really ore beds, frequently lenticular in shape, there is, to my mind, but little doubt.

The limonites (brown oxides), particularly plentiful in our ore deposits, are undoubtedly the result of the transformation or disintegration of other species of iron bearing strata, occupying similar positions, not capable of a clear definition, and yet undoubtedly related to the accepted parental deposits.

GEOLOGIC AND TOPOGRAPHIC RESUME.

In nearly every one of the accepted geologic positions, either large or small deposits of iron ores are to be found. Confined to but few of these geologic positions, however, are deposits in character and extent sufficient to warrant their development. To two ages, viz., the archean and cretaceous, it is generally accepted, belong the iron ores of this state. One of the accepted and most prolific of these is termed the archean, it carrying the highest of the grades of iron ores, the magnetites and the hematites. From the archean age comes the most extensively worked of our iron ores, it furnishing the major portion of the iron used in the union. Its area embraces the fields of the Lake Superior region, northeastern New York, and southeastern Canada, northwestern New Jersey, southeastern Missouri, Minnesota and Washington; the latter almost, as I have shown, in its entirety.

In that geologic horizon, yclept the cretaceous age, in which the majority of our coal beds are found, occurs also a considerable number of our beds of iron ore. These comprise oölitic hematite (a red oxide of iron of granular surface), limonite (a brown hematite), bog ore, brown clay, ironstone and black band.

ECONOMIC ADVANTAGES.

Prime factors to the successful operation of iron ore beds are generally accepted to be proximity to the fuels and fluxes requisite in the work of reducing successfully the ore to the metallic state; the freedom of the ore from ingredients precluding, even through process of smelting, their expulsion; and last but not least, the yield of metal of which the ore is capable. Phosphorus and sulphur are the main obstacles in the way of the successful manufacture of iron ore into steel. I shall endeavor, further on in this chapter, by analytical demonstration, to prove that our iron ores, the ages and varieties of which I have already noted, are to a degree free from the injurious ingredients named. A leading economic advantage, I submit, will then be illustrated.

AS TO FUELS AND FLUXES.

In a preceding chapter I at considerable length treated of Washington fuels. Therein are noted coals of the lignite, bituminous and coking classes, with both local analytical and comparative analytical statements.

The fuels best adapted for the reduction of the iron ores found in Washington to the metallic state, are dry-burning bituminous coals, coke and charcoal. Limestone is accepted in all iron ore treatments as a flux.

I have demonstrated the vast extent of our coal measures. I have shown that the majority of our bituminous coal seams in process of mining produce coals capable of being manufactured into coke of superior quality. Briefly let me chronicle the fact that contiguous to, almost at the very doors, so to speak, of our iron ore deposits, are located measures of coals unsurpassed as fuel necessary to the successful reduction of iron ores; and an idea may be had of the truly unexcelled advantages the iron manufacturer may enjoy in the pursuance of his calling in Washington.

To deposits of limestone, which is the necessary flux in the work of reduction, I have referred in another chapter, but will here repeat that they are at once of practically limitless extent, and like the necessary fuel, to be found in such close proximity to our iron ore beds as to reduce expense incident to their removal, etc., to a minimum.

CHARACTER OF IRON ORES.

In the prefatory chapter I alluded to the ages in which our iron deposits were formed; to the general character (as regards classification) of these ores; to the vast area in which are located iron ore deposits. In this chapter I shall confine myself to a description of these ore beds, and a detailed *résumé* of their product.

The whole of the great basin of Puget Sound may be said to comprise successive strata of coal and iron. To confine by topographical boundaries the extent or area of our iron ore beds would be impossible, for no accurate survey of their extent has ever been made. The statement hereinbefore made that in fourteen counties in the great basin, iron ore deposits of greater or less extent are to be found, must then suffice.

East of that axis known as the Cascade range of mountains, the deposits, while not so extensive as those found in the western part of Washington alluded to, are found existent in eight of the counties comprising the political subdivisions of that half of the state.

The principal products of the beds in both portions of the state may be noted as:

Siderite, or spathic or carbonate iron ore.
Clay ironstone.
Kidney ore.
Black band.
Magnetite (magnetic iron ore).
Hematite (red oxide of iron).
Specular iron ore.
Limonite (brown oxide of iron).
Bog ore.

LOCATION OF DEPOSITS.

To more particularly designate the location of deposits of the varieties of ore mentioned in the preceding chapter, I will state that the limonite variety is found along the Skagit river; the specular, up about the headwaters of the Snoqualmie, Skagit and Stillaguamish rivers; the magnetite, in the Cascade mountains at altitudes of from 800 to 2,500 feet above the principal streams having their source on the western slope of that range, and about the headwaters of the Cle-Elum, Teanaway and Yakima rivers on its eastern slope; the black band, kidney and carbonate ores, mostly in Whatcom county along the Nooksack river, in Skagit county along the Skagit river, in Snohomish county along the Stillaguamish river; and the kidney and carbonate in Stevens county along the line of the Spokane Northern Railway, and in Whitman county near Rock creek. The bog ores are found underlying the low lands which skirt the shores of Puget Sound.

All the varieties named, except the bog ore and magnetite species, are also found in the other counties named in the introductory chapter. In localities I have not particularized, such small measure of development has been accorded the deposits as to preclude extended mention of them.

Of the properties located as iron ore claims I am possessed of but little data, the amount of development work completed upon them being uniformly small.

SKAGIT RIVER BEDS.

The development completed here is that to be seen on the several deposits, styled by their owners "veins," of J. J. Connor & Co. This firm's property comprises four locations on what it has named the Tacoma vein, which is twenty feet in width. The locations are on the south side of Skagit river, about six miles north of Hamilton, Skagit county, at an altitude of 800 feet above the the river bed. The trend of the vein is northwest and southeast. The ore is a specular iron ore of good quality.

The Last Chance, a location of the same parties, is a vein ten feet in width, of the specular variety. The altitude of this vein is 400 feet, and it is located 700 feet south of the Tacoma vein.

The Washington is still another vein belonging to the same firm. The Mabel vein parallels the Last Chance, and has been traced on the surface for a distance of 2,000 feet. The ore is specular and hematite, and the vein is fourteen feet wide.

One mile south of the Washington is located the Tyee vein, traceable for three miles, and varying in width from sixteen to seventy-five feet. The ore carried is a brown hematite. This vein is also the property of Messrs. Connor & Co.

THE O'TOOLE CLAIMS,

The merit of the products of which have gained for them much celebrity, are located along Skagit river, to the northward of Connor & Co.'s properties. The group comprises ten locations. The veins vary in thickness from fifteen to sixty feet. The ores are of the specular, red and brown hematite varieties, and a black mottled ore, a mixture of magnetite and silicate of iron. These claims are being steadily developed.

THE SEATTLE MANGANESE IRON COMPANY

Is the owner of nine veins of iron ore, numbering no less than twenty-five locations. These veins lie opposite the town of Birds View, on the south side of Skagit river, in Skagit county. The veins vary in width from twenty to seventy-five feet. Improvement comprises a tunnel through ore on one of the locations called the Katie. This tunnel has been driven a distance of twenty feet. On the Eagle, another claim, a tunnel nine feet in length has been driven through ore. Only assessment work, in the shape of trails, roads, etc., has been done on the other claims. The ore is a specular and silicate of iron, carrying manganese. The gentlemen comprising the company are Hon. C. H. Hanford, federal judge; Hon. John H. McGraw, J. W. McLeod, Frank Hanford, Fred Sander, A. E. Hanford, and Herbert Tappan.

THE DENNY MINE

Is the oldest location in the State, and shows the most development work. This mine is located on a peak of the Cascade range of mountains to the westward of Mount Logan, about two and a half miles from the Seattle, Lake Shore & Eastern Railway, in King county. The mine is accessible from this point, and also by a route through a small valley. The Denny vein is one of the best deposits of magnetite and specular iron ore in Washington. What is known both as the Chair Peak and Kelly claim represents a deposit, possibly, of the largest and purest of all the deposits of magnetite ore yet discovered in the state. This claim is located some six miles from the Denny mine.

A bed of ore on Mt. Logan, also of the magnetite variety, has been located as "the Guy claim" for several years, and is sometimes designated the Summit mine. Its elevation is estimated at 1,250 feet above the Seattle, Lake Shore & Eastern Railway, from which it is distant not more than ten miles.

IN KITTITAS COUNTY,

Not far distant from the summit of the Cascade range, on its eastern flank and lying along the Cle-Elum river, in Cle-Elum valley, are five very considerable beds of iron ore of the magnetite variety.

The first of these beds is found located on the east fork of the Cle-Elum river, about seven miles above Cle-Elum lake, and twenty-five miles distant from the main trunk of the Northern Pacific Railway system. This ledge has been traced and located for a distance of over two miles; it is well defined, and has a trend toward north and south. In width it varies from thirty-five to seventy feet. The pitch is toward the west, the angle being at almost twenty-two degrees. The case rock is a porphyry. Four more separate beds or veins comprise this deposit, which is the property of the Pacific Improvement Company.

West of the town of Cle-Elum, in Kittitas county, is the Nelson Brothers property, comprising the Iron Queen, Iron King, Grand

Center, Union, Twin and Low Point claims. The ore found is of the brown hematite variety.

South of Cle-Elum are located the properties of Thomas Johnson, of Seattle, and of the Ellensburgh Improvement Company. Mr. Johnson calls his claim the Grand Entry. The ore is a brown hematite, as is also that found in the claims of the Ellensburgh Improvement Company, which are the Iron Heart, Iron Crest, Nigger Baby, Uncle Sam, and Brother Jonathan locations. Some valuable magnetite properties contiguous to those of the Pacific Improvement Company, are owned by Mr. Frank C. Ross, of Tacoma, and Moses Emerson, of Ellensburgh.

IRON ANALYSES.

As I have stated, the measure of development accorded all the iron ore claims of this region is uniformly small. Because of this fact convincing data regarding the value of the product remains to be gleaned from the result of analytical demonstration. I have made numerous assays of all the species here existent, and am in possession of the results obtained by other analytical chemists. I append them.

SKAGIT RIVER ORES.

Mine.	*Met. Iron.*	*Silica.*	*Sulphur.*	*Phosphorus.*
Connor & Company's mines:				
Tacoma vein	58.13	4.12	.19	.06
Last Chance vein	60.00	7.03	.04	trace
Washington vein	56.72	5.94	.03	.07
Tyee vein	63.07	7.03	trace	trace
The O'Toole claims:				
Claim No. 1.	57.28	7.70	.17	.09
Claim No. 2	56.72	10.13	.16	.02
Claim No. 3				
Claim No. 4	57.17	7.19	.17	.07
Claim No. 5				
Claim No. 6	54.23	12.13	.13	.03
Claim No. 7.				
Claim No. 8	50.90	13.29	.12	.01
Claim No. 9	50.29	9.50	.19	.10
Claim No. 10	54.61	8.47	.15	.07

BESSEMER STEEL ORES.

Mine.	*Met. Iron*	*Manganese*	*Silica*	*Sulphur*	*Phosphorus*
Seattle Manganese Iron Co.'s Claims:					
Katie vein	62.00	3.00	6.50	trace	trace
Eagle vein	58.00	4.13	7.00	trace	trace
Vein (Tennessee) No. 3	42.90	2.00	21.93	.00	n. d.
Vein (Puget) No. 4	56.38	5.67	12.47	.00	n. d.
Vein No. 5	48.00	n. d.	13.19	trace	n. d.
Vein No. 6	47.30	n. d.	13.11	trace	n. d.
Vein No. 7	54.03	n. d.	9.40	trace	n. d.
Vein No. 8	57.20	n. d.	6.92	.00	n. d.
Vein No. 9	52.09	n. d.	13.24	.00	n. d.

KING COUNTY ORES.

Mine.	*Met. Iron*	*Silica*	*Sulphur*	*Phosphorus*
Denny Vein	62.13	7.01	.03	.03
Chair Peak Vein	63.90	6.13	.04	.03
Mt. Logan	67.12	3.62	.05	.02
Green River	45.51	10.12	·05	trace

KITTITAS COUNTY ORES.

Mine.	*Met. Iron*	*Silica*	*Sulphur*	*Phosphorus*
Pacific Improvement Co.'s Vein	56.50	9.10	.08	.18
Nelson Bros	48.11	8.07	.10	.11
Frank C. Ross	66.50	5.10	.08	trace
Ellensburgh Improvement Co., No. 1	55.00	7.90	.13	.13
Ellensburgh Improvement Co., No. 2	54.30	8.02	.09	.18
Moses Emerson Vein	65.00	6.05	.04	.20

WHATCOM COUNTY ORES.

Mine.	Met. Iron...	Silica.......	Sulphur.....	Phos-phorus....
Prospect in northwestern part Whatcom county, on the Nooksack river..........	59.27	9.97	trace	trace

BOG ORE ANALYSES.

Mine.	Met. Iron...	Silica.......	Sulphur.....	Phos-phorus....
From deposit on Bellingham Bay, Whatcom county,..	46.20	7.12	trace	.09

COMPARATIVE ANALYSES.

Herein find comparative analyses of iron ores, offered that the value for steel purposes of our iron ores may be the more accurately ascertained.

A TYPICAL BESSEMER STEEL ORE.

1. Metallic iron.................... 61.00 per cent.
2. Silica............................... 6.13 per cent.
3. Sulphur............................... .17 per cent.
4. Phosphorus......................... .04 per cent.

STEEL MAKING IRON ORES.

Mine.	*Met. Iron...*	*Silica.......*	*Sulphur....*	*Phos-phorus....*
Lake Superior...	67.13	3.05	trace	.05
Iron Mountain, Missouri...	64.90	5.12	.01	.04
Tennessee..	62.81	5.10	.03	.02

MINES AND MINERALS OF WASHINGTON.

PART FOURTH.

SECOND ANNUAL REPORT OF GEORGE A. BETHUNE,
STATE GEOLOGIST.

WASHINGTON LIMESTONES.

Possibly no integral factor of the minor order will be found to exert a greater influence in the extensive development of a deposit of anything capable of being manufactured into a commercial article of value, than a component requisite to such manufacture.

In the previous chapter, which is one on the iron ore deposits of Washington, I alluded to the great intrinsic value of limestone as a flux, requisite in the process of reduction of such ore to the metallic state. Taking into consideration that statement, together with the one also recorded, that in close proximity to our iron ore deposits were to be found extensive deposits of limestone, and the desire of the writer to treat *seriatim* the different subjects which to my mind should find place in this volume, I deem it here appropriate to refer briefly to Washington limestones, prefacing my remarks in this connection with a brief allegoric prelude.

Limestones may be classified into two divisions, which are the calcite and dolomite.

The calcite variety is popularly termed a carbonate of lime, and the dolomite a magnesian limestone.

The calcite has no definite determination as regards its color. Calcite limestone is found in gray, red and drab hues frequently, these different colors being ascribed to the presence of various impurities in the rock. The coloring of the calcite limestone to a reddish hue, for instance, is due to the presence of iron in the rock. Carbonaceous matter contained in the rock causes it to assume a drab color; and the gray color is due to the presence of manganese. When calcite limestone is of the first named color, it is known as a ferruginous limestone, and this latter variety is found in large quantities in proximity to our iron ore deposits.

The calcite is termed a "hot lime" when burned, cooled and slacked, and the dolomite a "cold lime" after having undergone the same treatment.

As is the case with the best architectural sandstones, the best limestones are found in the region bordering on the archean. Both calcites and dolomites are to be found in Washington. On several of the islands of Puget Sound calcite limestone deposits of extended area are being worked with profit, notably, on the island of San Juan, a member of the Puget Sound archipelago, where for several years these deposits have been mined and the product resolved into a lime of fine quality and great repute.

As stated, associated with those of our iron ore deposits in the Cascade mountains, are extensive deposits of limestones of the calcite and dolomite varieties. The deposits may be easily quarried, and the product has been found to be of fine quality, and valuable at once for architectural and for fluxing purposes. Here the calcites are found in different hues, including blue, white, drab, and red. Sometimes we find them, as regards their composition, very fine grained; and sometimes, by the process of crystallization, transformed into marbles of the most true and beautiful character. Especially in King, Stevens and Kittitas counties are these deposits found very extensive. Identified, by reason of their contiguity to iron ore deposits there found, are limestone deposits of both the calcitic and dolomitic varieties, along the Skagit, Nooksack, Skykomish, Stillaguamish, Sauk, Snoqualmie, White, Green, Puyallup, Nesqually, Cowlitz, Yakima, Okanogan, Methow and Colville rivers. I mention this fact that a better idea may be had of the unparalleled advantages which may be enjoyed in the development of our extensive iron ore deposits, and also the reduction of these ores to the metallic and merchantable state.

Along all the streams I have mentioned, and on many in locations contiguous to marts of trade, deposits of limestone eminently well adapted to the uses of architecture abound.

A deposit of what may be termed "hydraulic lime" attracted my attention on the occasion of a visit paid Okanogan county the past summer. This deposit is located in Galena mining district, in the eastern portion of the county, in what is known as the "lime belt." A large amount of clay and silicious matter I found intermixed with the lime, which, after being burned, refused to slack with water, the presence of the clay and silicious matter preventing it from so doing. When ground and mixed into mortar, however, I found that it set exceedingly hard under water.

A soft, earthy limestone, which is really chalk, is found along the Columbia river, near the confluence with that stream of the

Wenatchee river. This chalk is almost pure white in color, and should be valuable for manufacturing into crayons for marking purposes; and, intermixed with certain of our clays, might be resolved into a merchantable quality of cement.

About the only extended use now being made of our limestones is confined to San Juan island, mentioned above. Here the kilns of the Tacoma & Roche Harbor Lime Company have been in operation for several years, the output of lime (which is accepted as of standard quality) aggregating at times 1,600 barrels per diem.

TABLE OE LIME ANALYSES.

Elements.	*Island Co...*	*Okanogan...*	*King.........*	*Skagit......*	*Cle-Elum...*	*Stevens.....*	*Kittitas.....*
Carbon of lime..............................	98.21	97.13	98.34	96.50	97.50	98.00	95.30
Silica..	.04	2.30	1.08	2.01	1.05	1.25	2.70
Maganese....................................	1.01	....	....	....	....	....	....
Iron...	....	....	....	.19	.61	....	1.10
Magnesia....................................	....	.10	....	....	....	....	....
Phosphates................................	.01	....	....	....	....	....	....
Water...	.73	.47	.58	1.30	.84	.75	.90
Carbon, organic..........................	....	....	....	....	....	....	....
Totals....................................	100.00	100.00	100.00	100.00	100.08	100.00	100.00

WASHINGTON MARBLES.

In a preceding chapter, treating of our building stones, I have alluded to the existence of marble deposits in this state. In my first annual report (Mines and Minerals of Washington, 1890), I briefly alluded to these deposits, reciting the fact that their development had then been conducted on a very meagre scale. The same may be said of the development accorded these deposits during the past twelve months.

It is a matter of some surprise to me that a greater share of public attention has not been paid the matter of rendering at once productive and profitable these deposits. Marble, of recent years, has entered very largely into use, not alone as a component part of the manufacture of furniture, and the ornate decoration of architectural structures, but as a leading factor in building construction. In fact, marble may be accepted as the popular "fad" with builders of the most magnificent, and hence most costly, structures of the time.

In Washington marbles should certainly play a prominent part in building operations. Washington marbles have been found eminently well adapted to such usage; and the fact of the wondrous beauty of some of the varieties found here, coupled with their demonstrated utility for building purposes, seems to me to warrant, on a far more extended scale, their introduction into our building construction. I can say here, and without fear of contradiction, that as handsome and durable structures of finest marble (of home production) may adorn the streets of our cities as can be found in any municipality in the world; if development is given our deposits, and they in consequence are made productive.

GEOLOGICAL.

Any limestone susceptible of a high polish is generally termed a marble. Limestones of the crystalline character mentioned in the chapter on these stones furnish the largest proportion of marbles.

The verd-antique ophiolite varieties, the most beautiful and valuable for sculptural purposes, owe their beauty and incident value to certain minerals disseminated through the limestone forming mixtures. I have already alluded to the adaptation of some of our marbles to purposes of ornamentation. As stated, marbles best adapted to this purpose are those posessed of fine even texture, free from foreign mineral substances, and of a pure and uniformly white color. While it is the popularly accepted idea that such marbles are rare, and while popular demand in this regard looks to the deposits of Carrara and Parros for supply, let me inform you that marbles the equal, if not the superior of these, are to be found in abundance right here in our own state.

Marbles best adapted for building purposes are here also in abundance; and are not found contaminated with mineral substances to the extent of deposits found elsewhere, and are not too coarsely crystalline—a grievous fault. Many of our building marbles are streaked with serpentine, enhancing much the attractiveness of their general appearance. Dolomitic marbles are also found in several deposits. Of the crystalline species exceedingly fine-grained and homogenous varieties are to be fonnd. These latter have been elsewhere esteemed most valuable for building purposes. Again we find here what is called a marble, but which is really a limestone, of a very compact character and of variegated hüe, which is susceptible of a high polish, and which should be turned to profit, as it certainly would serve as ornamental stone.

DISTRIBUTION OF DEPOSITS.

Marble deposits and beds of compact limestone of variegated hue and susceptible to a high polish (as just described) are widely distributed in Washington. They may be found along the Skagit river and its tributaries, in Skagit county; along the Columbia river (Washington side), and in Cowlitz, Okanogan and Stevens counties. In Stevens county is a quarry of grayish black marble of great beauty. Therein has been placed the necessary plant for quarrying the deposit. Many samples obtained from these deposits and subjected to analytical tests have demonstrated their fitness for use in both the fine arts and for building purposes. I will here describe the method adopted in analyzing samples of the marbles to which I have referred.

First is made a microscopical examination of the marble. From it I pronounce it to be hard and very strong. My samples, when subjected to the treatment, took splendid polish, assuming the smoothness of glass shortly after the application of ordinary marble polishing materials. Duller granules, which I have heard complained of in this stone after attempts at polishing it, I find can be dismissed if the polishing process has been done in a workmanlike manner known to artisans used to this work.

Next I put my samples under chemical treatment. As a result I found that they contained about 7.80 per cent. of carbonate of magnesia, and another microscopical examination demonstrated the fact that two different kinds of granules made up the stone, with different degrees of hardness.

This observation, taken together with the fact that a residue remained after putting the marble under treatment with cold dilute hydro-chloric acid, which residue was easily soluble in hot hydrochloric acid, established the fact beyond peradventure that the Stevens county marble is a dolomitic limestone, and consists mechanically of a mixture of calcite granules (carbonate of lime) and dolomite granules (double carbonate of lime and magnesia). The mechanical mixture of the two kinds of granules, by scratching the surface and applying a drop of cold dilute hydrochloric acid, could be plainly seen by the naked eye. The calcite granules are left in relief. A further residue, insoluble, in the acid exists. I find the cause of the clouding of what would otherwise be a marble of snow-like whiteness to be due to graphite scales. The greater number and greater proximity to one another of the graphite scales, the darker the coloring. Graphite scales very thickly scattered through the cleavage of the grayish-black variety are accountable for the mottled appearance of that stone. I have seen in the purest white marble small scales of graphite, which, without the aid of the microscope, would have remained undiscovered. The specific gravity of the samples, and other interesting information concerning these tests, I append: Specific gravity, 2.69; weight of one (1) cubic foot, 168.12 pounds; the proportion of water is .15; the loss in carbonic acid solution, after long exposure, amounted to .94. When I heated my samples in the muffle, aside from an increased whiteness, like that characteristic of limestone under the same treatment, I could observe no change in the marble. When under the effect of a full red heat I noted the appearance of the marble changed to a mottled color, and shortly afterward was not surprised to see the marble act identically with limestone under like conditions. That is, it began to crack on the surface and edges, but the cracks penetrated but a little distance into the interior of the sample. Withdrawing the sample from the muffle I immersed it immediately in cold water. The result was simply a corrosion of the edges and cracked surface, thus clearly establishing in my mind the general strength and worthiness of the marble for architectural purposes. I next determined, with the aid of a government machine, the crushing strength, using a bed cube 1.525 x 1.540 inches (h. t.), giving 16,992 pounds per square inch; and on an edge cube 1.369 x 1.339 x 1.420 inches (h. t.), giving 16,102 pounds per square inch.

I think these returns a practical demonstration of the value of the marble.

WASHINGTON SANDSTONE.

In my first annual report (Mines and Minerals of Washington, 1890) I at some length treated of the deposits of sandstone existent in Washington, of a character rendering possible their adaptation to structural usage. In that report I referred to the fact that the architectural era seemed to be of the stone age. During the past twelve months this fact has been even more strongly impressed upon my mind than before. I see in the vast and rapidly increasing quantities of suitable building stones demanded additional evidence of the fact. It may be appropriately here chronicled that the demand for stone for architectural purposes has increased three fold in the past year, and there is no prospective cessation of the demand. Stone has entered more largely than ever into structural usage, and its demand even as part of the structural material for private residences has assumed extended proportions. We have come now to deem it positively indispensable to complete construction as regards our buildings erected to suit the purposes of commercial character; without introducing sandstone in such construction; it is also an indespensable adjunct to the general ensemble requisite in caravansary construction. In short, in nearly every architectural project, stone seems to be playing, and bids fair hereafter indefinitely to play, next to iron, a most prominent part.

I do not wonder at this. No such material, from the time of the introduction of the higher order of architecture as applied to buildings adapted to all purposes, by the Greeks, has been considered such an essential requisite to strength, durability and ornateness in building as stone. When it is considered that, here in richly endowed Washington, are to be found stones capable of use in architecture, and not excelled in this regard or in susceptibility to beautification at the hands of the skilled artisan by any other known, and that the expense of their procurement and moulding may be said to be minimized to a degree, it should not be a matter of surprise that they have so largely entered into our architecture. Chief among the stones of Washington, valuable for building purposes is sandstone.

The sandstones of this state comprise the argillaceous and silicious varieties.

Sandstone may properly be termed simply the consolidation of sand, brought about by pressure. Consolidation of the sand is rendered permanent with the aid of small quantities of clay adhering to the grains of the sand. Such consolidation, so brought about, is resultant in the formation of that species of sandstone known as argillaceous.

The silicious species is formed also by consolidation of sand held together with silica acting as a cement.

The sandstone found in this state is of varied hue. We have the stone in colors of red, drab, buff, yellow, olive color, and brown. The colorings as described are traceable in their origin to mixtures of iron oxide with clay in the composition of the stone. These ingredients may also be said to act in the capacity of cements, thus perpetuating the consolidation of the sand. Sandstone deposits of both the argillaceous and silicious varieties, capable of use in architecture, are widely distributed over the area comprising Washington. They are to be found in Whatcom (in the vicinity of Bellingham Bay), Skagit, Snohomish, King, Pierce, Thurston, Lewis, Chehalis, Cowlitz, Clark, Clallam, Kitsap, Mason and Jefferson counties in Western Washington; and in Kittitas and Stevens counties in Eastern Washington. In Whatcom, Thurston, Pierce and King counties several extensive quarries have for some time past been in successful operation.

AN ANALYTICAL TEST.

I have been given an opportunity of analyzing a sample of sandstone (olive green in color) from the Chuckanut quarries, Whatcom county. The method employed I will describe.

A rough cubical specimen of this sandstone, weighing about 100 grammes, I dried at 100 c. c. to a constant weight five days; weighed in the air, and then weighed after prolonged immersion in water, suspended by a horse hair. The weight of the dried cube, divided by the loss of weight caused by its immersion in water, was taken as its specific gravity. One cubic foot of this sandstone's weight I next calculated from the weight of a cubic foot of water, which is equal to the weight of 62.4 pounds. My next effort was directed toward a determination of the crushing strength of the rock. I used a government machine of 100,000 pounds. The absorption of moisture was next determined. My determination of the absorption of moisture was successfully conducted by carefully drying roughly cubical fragments in a water vapor saturated atmosphere for fifty-one days, these cubes being of 100 grammes weight. The cubes I placed under a jar over placid water and on a shelf of glass. For a second test I took an equal weight of the samples and placed them in a vessel of water for five days. After immersion for the above length of time they were dried carefully, and their weight thereafter represented the amount of moisture absorbed by them during their immersion. To ascertain the carbonic action on the stone, I drowned a similar weight (100 grammes) in water in a carbonic acid solution.

After this treatment I found that the loss of weight was inconsiderable. It was a practical demonstration to me of the worth of our sandstones. But, desirous of irretrievably establishing the quality of the stone, I determined to ascertain the weathering and the staining effects that could be developed within the period (fifty-one days) during which I gave the samples their airing. This was to approximate what could reasonably be anticipated of our sandstones used for building purposes, as regarded the maintenance of their natural color. I noted the effect of a dry heat upon my samples by placing the cubes in my muffle furnace and subjecting them to a red heat. Raising the heat, I gradually made it a full redness. I noted no softening or cracking after they had undergone this ordeal. I plunged them while subjected to this heat in a bath of cold water, and the only perceptible change was a slight corroding of the surface.

ANALYSES OF SANDSTONE.

Specific gravity	2.570
Weight per cubic foot	160.620
Absorption, moisture	1.350
Absorption, water	5.342
Loss on exposure to carbonic acid gas solution	.180

Exposed to acid fume, the stone was stained a greyish yellow in spots. It became loosely adherent on the surface and through quiet integration lost 6.13 per cent. of its weight, and by friction an additional loss was incurred of 10 per cent. I found after treatment in my muffle furnace a change of color, from the green to a greyish hue. I could not, after giving the samples a full red heat, develop a single flaw, nor could I develop this after immersing them while hot in water, barring the corroding of the face of the samples to which I previously alluded. In determining the strength that would crush these samples I used cubes 1.715 x 1.755 x 1.650 inches (h. t.) for bed, and 1.390 x 1.150 inches (h. t.) for edge, resulting in 10.321 pounds per square inch for bed, and 8.100 pounds per square inch for edge.

WASHINGTON GRANITES.

Considering the fact that deposits of granite may be said to abound in the state, it is a matter of some surprise to me that this stone, universally deemed among the most valuable and ornate of building stones, has not more largely entered into local architectural use. We certainly have among our granites the most beautiful to be found, while as regards quality and adaptation for structural use, none of the granite here found can anywhere be excelled.

The Washington granites belong chiefly to the archean or azoic ages, and are composed of quartz, feldspar (orthocdase), hornblende and mica. While some of the granites are here found coarsely crystalline and carrying iron pyrites, thus rendering them pervious to water, and hence not valuable for building purposes, other granites of Washington are found free of these injurious substances, and eminently well adapted to the purposes of architectural usages. These latter are found to be of close texture, and containing quartz, feldspar, hornblende and mica only in small quantities. This variety is found by workmen to split at right angles ("the rift") very readily, and to be easily dressed into shapes desired.

The Washington granites, as far as color is concerned, vary. Some are found of a dark whitish gray, reddish and gray color. The dark coloring is given the granite by the hornblende it contains. The whitish gray and gray hues are given the granite by the light colored feldspar contained, and the reddish color by red feldspar. What is popularly here termed granite is nothing more nor less than syenitic granite, and deposits of this composition underlie all the coal rocks lying to the westward of the Cascade range of mountains in the Puget Sound basin, because syenitic granite really forms the core of that range. It is found at lofty altitudes along these mountains.

Excellent architectural granite is also found west of the Cascade range, along the Skykomish river, Green river, and Stillaguamish river.

On the eastern side of the mountains, a like quality is to be found along the Cle-Elum river, and at a point distant a few miles south of the city of Spokane.

WASHINGTON CLAYS.

Washington is rich in the possession of the various clays adapted to manufacturing purposes, the deposits being as widely distributed as the varieties found are numerous. There are here to be found, as stated in my first annual report, clays adapted to the manufacture of brick, both rough and pressed.

Clays adapted to the manufacture of fire brick of fine quality.

Clays adapted to the manufacture of pottery of all descriptions.

Clays suitable for the manufacture of terra cotta and other essentials to ornateness in buildings, etc.

Clays capable of being formed into porcelain and manufactured into wares the equal in quality and appearance of any imported from abroad.

Clays adapted to the manufacture of all kinds of stoneware.

GEOLOGICAL.

The origination of clay has been traced with a sufficient degree of accuracy to warrant the statement that it comes from feldspathic rocks, such as granites, porphyries, etc. All clays are essentially silicate of alumina.

They are found, as regards color, to vary, and in this state clays of many colors are found. The pure clays are white, while those of other hues are the result of the admixture of foreign substances contained within them. In Washington clays are found to occur in both the tertiary and cretaceous ages, and the deposits are located along the borders of Puget Sound and on the islands dotting the surface of the Sound waters.

Clays adapted to manufacturing purposes of different characters are also to be found underlying the coal strata of the great Puget Sound basin. By proper weathering these particular clays may be rendered highly valuable for manufacturing into fire brick, pressed brick, rough brick, terra cotta, drain pipe, etc.

PROPERTIES OF CLAY.

Clays possessing to a high degree plasticity, and because of this termed generally "fat" clays, are of most value to the potter.

These clays, properly moistened, are found possessed of great tenacity, are pasty, and hence easily formed into any shape the potter desires. Clays possessed of no great amount of plasticity, and when worked, even with the aid of moisture, have but little of the tenacious and pasty character of "fat" clays, we term "short" clays. Both varieties mentioned have been used in the manufacture of building brick in this state for several years, the product, as a rule, being of most excellent quality. An essential requisite in all clays deemed suitable for purposes of manufacture is a sufficient proportion of "true" clay or kaolin.

"True" clay or kaolin is an unctuous, hydrous silicate of alumina, usually white in color, and generally in its component aggregate containing 14 per cent. water, 40 per cent. alumina and 46 per cent. silica. In many of the best manufacturing clays found in Washington kaolin enters largely into the general composition, but minor proportions of other substances being traceable. While constituting singly a clay of unexcelled quality, kaolin is the prime essential ingredient of all true clays. An admixture of oxide, carbonate or sulphide of iron in clay makes it to the potter of but little value, since, when found to the extent of 2 or 3 per cent. in a clay, they impart a yellow, red or brown color, according to the amount of either contained, or the degree of heat to which the moulded article is subjected. In the Washington clays adapted to the manufacture of pottery but little of these contaminating substances are to be found.

Essential ingredients to good brick clays are kaolin (in sufficient proportion) and quartz sand. The kaolin in proper proportion is requisite that the necessary adhesiveness and plasticity may exist, and the quartz sand to prevent the tendency of the clay when burned to shrink, warp and crack. Mingled with these in brick clay are to be found iron oxide, potash, soda, lime and magnesia. It is the presence in clay of iron oxide that gives to the brick, after kilning, its common red color, the oxide becoming red oxide.

The bricks seen in some of our principal business blocks which are of a rich cream color, as in the structure of the Fidelity Trust Company's block, Tacoma, owe that hue to the presence in the clay from which they were manufactured of a considerable proportion of lime and magnesia, or of these and potash, which at a high temperature have formed a substance with the iron and silica contained in the clay, which has partly fused, thus giving to the brick a greater degree of solidity.

Our Washington brick clays will be found to improve much on weathering.

ANALYSES OF FIRE CLAY.

CLAY MINE, KING COUNTY.

Silica	57.50	Magnesia	1.00
Alumina	34.37	Alkalies	.68
Iron oxide	1.24	Waters	4.71
Lime	.50	Total	100.00

This clay stood, before fusing, a temperature of 2,760 degrees F.

MACINTOSH BED, GREEN RIVER, KING COUNTY.

Silica	69.71	Potash	.19
Alumina	18.39	Soda	.83
Iron oxide	1.44	Lost by ignition	8.94
Lime	.35		
Magnesia	.15	Total	100.00

Fire test, 2,831 degrees F.

CUMBERLAND CO. (CONNOR MINES), SKAGIT COUNTY.

Silica	49.73	Soda	1.10
Alumina	32.57	Potash	.00
Iron oxide	1.99	Water	12.38
Lime	.95		
Magnesia	1.28	Total	100.00

ANALYSES OF POTTERY CLAY.

BELLINGHAM BAY, WHATCOM COUNTY.

Constituent	Per cent	Group	Total
Combined silica	43.30	Kaolin constituents	94.41
Alumina	35.21		
Water	15.90		
Titanic acid	1.11	Inert constituents	2.31
Sand	1.20		
Lime	.08	Fluxing constituents	3.11
Iron oxide	1.07		
Magnesia	.14		
Potash	.52		
Soda	1.30		
Maganese	.00		

PORT ANGELES, JEFFERSON COUNTY.

Constituent	Per cent	Group	Total
Combined silica	35.40	Kaolin constituents	80.50
Alumina	31.20		
Water	13.90		
Sand	10.19	Inert constituents	10.19
Titanic acid	.00		
Lime	.44	Fluxing constituents	2.65
Iron oxide	.70		
Magnesia	.18		
Potash	.52		
Soda	.36		
Manganese	.45		

GREEN RIVER, KING COUNTY.

Constituent	Per cent	Group	Total
Combined silica	39.72	Kaolin constituents	78.07
Alumina	27.17		
Water	11.18		
Sand	16.03	Inert constituents	16.03
Titanic acid	.00		
Lime	1.12	Fluxing constituents	3.87
Iron oxide	1.54		
Magnesia	.00		
Potash	1.21		
Soda	.00		
Manganese	.00		

A TYPICAL FIRE CLAY.

(Golden, Colorado.)

Silica	53.20	Soda	.00
Alumina	31.01	Potash	.31
Iron oxide	.66	Water	14.46
Lime	.20		
Magnesia	.16	Total	100.00

A TYPICAL POTTERY CLAY.

(Middlesex county, New Jersey.)

Combined silica	46.11	Kaolin constituents	96.56
Alumina	35.56		
Water	14.89		
Sand	.51	Inert constituents	1.97
Titanic acid	1.46		
Iron oxide	1.38	Fluxing constituents	1.50
Lime	.00		
Magnesia	.00		
Potash	.12		
Soda	.00		

THE SOILS OF WASHINGTON.

Proverbially rich and productive are the soils of the State of Washington. Their natural products are as varied as is the state's ensemble of resources. Note the result of natural arboriculture, floriculture and horticulture throughout the great state, and an idea may be gained of what may be resultant from the exercise of the arts employed by the agriculturist of our day in rendering the soils of our state capable of supplying the actual sustaining power of life.

To the more lucidly treat of the soils of Washington, for there are to be found several species, it might be best, at the outset, to so classify them as to admit of their being separately taken up and described.

Admitted to known classification in the accepted vocabulary of soils, are five species in this state, viz.: The humus, alluvium, drift, loam and basalt soils.

A humus soil may be described as one containing only a due proportion of clay, which, because of this fact, renders it peaty or a turfy soil, capable of being warmed and dried, when properly drained, very rapidly. But this humus soil, because of carbonic and other acids, is likely to "sour," and is generally found to be deficient in the essential mineral elements of plant food.

This soil, however, may be mixed with accepted fertilizers, the principal elements of which are warmth and freedom from the acids I have mentioned, with profitable results, especially when used in a humus soil largely deficient as regards its mineral elements of plant food. An essential requisite to the rendition of a humus soil cultivable is thorough drainage. The more complete the drainage the better adapted to the uses of mankind will be found our humus soils.

In the Puget Sound basin, that vast area lying between the natural axis of Washington (the Cascade range) and the waters of Puget Sound, the area is found to be almost wholly covered with vegetable mould, its presence being distinctly traceable to the heavy timber growth which from time immemorial has covered the surface. Where the growth is strongest, deepest is found the mould; where weakest are found the slightest evidences of it. Because of these facts the soils of the great basin, which I shall later on more particularly describe, must be classified as humus soils, not wholly because they are turfy or peaty, for in the upland sections of the basin they are not; not wholly because they must be drained or aided by fertilization to a state rendering them cultivable, for thousands of acres require the ministration of neither of these essentials; but because of a commingling, as regards the general character of these soils, of species of the humus variety; all, as an ensemble, making up a series of general characteristics, plainly calling for their classification as humus soils. It may be safely chronicled of the humus soils of the great basin, that they rank among the most valuable of the state's possessions, and are destined to play a most prominent part in its general upbuilding and advancement.

ALLUVIUM SOILS.

This title belongs to that soil which, in Western Washington, is found on the low-lying areas of the Puget Sound basin. Alluvium soils may be said to have been the transported sedimentary deposits of low-lying lands, swales and tide flats, all of which are prominent features of the general topographical aspect of the basin. No soil has been found more productive than has that termed Puget Sound basin alluvium; and it may be said to predominate over all other species of soil in this region. Enriched by both vegetable and animal matter, the soil of the tide flat areas is especially fertile; as note how productive have been and are the alluvium soils in Snohomish, Skagit and Whatcom counties. The various mountain streams finding source along the western flanks of the Cascade range, it should be noted, are bordered by but little alluvium soil.

DRIFT SOIL.

The rolling area lying to the northward and eastward of the humus soil deposits is found covered to a considerable depth with drift, bearing evidences of a long past acquaintanceship with both water and ice. In portions of this area, lying nearest deposits of both humus and alluvium soils, this drift deposit is found capable of being easily and profitably adapted to cultivation; and in the major portion of the drift-covered area are found farms yielding profitably of a great variety of agricultural products. The drift soils of Western Washington constitute principally the hill lands of the region. Sand, clay and gravel constitute the component parts of the soil. In some places gravel is so over-proportionate to the other substances mentioned that the land is rendered practically valueless for purposes of agriculture.

While really our drift soils should be called simply a glacial precipitate, it is a fact that in Western Washington they are found much more prolific than is generally the case with such precipitate. Western Washington drift soils have been found especially well adapted to the cultivation of most varieties of tree and vine fruits and the majority of edible vegetables.

LOAM SOIL.

Loam is a clay soil impregnated with fine-grained sands, rendering it mealy and crumbling in character. Loam owes its origination to the slates and sandstones, and its existence in Western Washington is first noted by the geologist on the more elevated lands and above the drift soils. It may be accepted as only of tolerable quality for purposes of cultivation, but as a producer of grasses its value has been thoroughly demonstrated. The area of loamy soil in Western Washington is larger than that covered by humus and drift soils combined. The altitudes at which it is found vary, its presence being noted on the low-lying hills bordering the western flanks of the Cascade range of mountains, and away up on the cretaceous hills and peaks of the range; in places, too, upon their very crests.

Loam soil is Western Washington's leading hay producer, and forms the principal portion of the grazing areas in this portion of the state. Left uncultivated, these soils, from the fecundating dust of red and white clover and the native grasses cast upon them, are known to grow these plants most luxuriantly.

BASALT SOILS.

With brief reference to the basalt soils of the state, I shall conclude. Let me, under this heading, state that soils of the humus, drift, and loam varieties are in Eastern Washington to be found wherever conditions described as explaining their presence in Western Washington are found similar.

Basalt soil is popularly accepted as the result of the decomposition of the basaltic rock. Such, however, is not entirely the case.

The metamorphic slates and other disintegrated rocks play, as regards the basaltic soil of Eastern Washington, a part in its general make-up.

In appearance, the basalt soil covering the vast area east of the Cascade mountains known as the great Columbia table-land, is of grayish color; in portions of the area, a dark gray; in others, black; in still others, light gray. In some sections I have found it compact, and evidence that clay constitutes part of its structure are found. Such condition, however, is rare. Of the soil, as regards its being a typical high producer, its structural character may be said to embrace a condition equally removed from too great compactness and too high a degree of permeability, which renders it a soil "capable of absorbing and retaining only a due amount of moisture, the while giving easy exit to any over-plus, to permit the ready access of air, and to absorb and practically utilize the warmth proper to its location."

The basalt soil in the great grain and fruit belts of Eastern Washington is found to contain proportions of phosphoric acid, lime, iron and organic matter sufficient to render it most prolific without the aid of fertilizers, to make it the most productive by far of all our soils when irrigated in places where natural moisture is deficient. The ingredients named above especially adapt a soil to the successful cultivation of the various cereals and fruits of both tree and vine.

Inequalities, as regards the productiveness of our basaltic soil, are to be found here as elsewhere; but these are not frequent. Where the out-cropping of basalt is noted, it will be very generally found that the soil thereabouts, while possibly rendered unworkable with the plow, may be rendered productive, hence profitable, with the aid of smaller and less cumbersome implements of agriculture.

In conclusion, let me add, that a soil possessed of the following may be found in any of the areas herein referred to; may in many be cultivated off-hand, and in few placed in readiness for cultivation with but slight labor and expense. These requisites and the quantities appended are accepted as essentials to productive soil: Easy penetrability to roots, to moisture, to air; sufficient retentiveness to prevent the too ready escape of moisture; readiness to at once absorb and utilize the warmth of the sun's rays.

ANALYSIS.

I append results of an analysis of soils taken respectively from the fruit belts of Yakima and Walla Walla counties, in Eastern Washington.

Constituents.	*Yakima*	*Walla Walla*
Silica	72.10	70.07
Alumina	14.08	12.01
Phosphoric acid	.10	.17
Sulphuric acid	.07	.04
Peroxide of iron	8.37	11.19
Lime	.44	.73
Magnesia	.95	1.03
Soda	.15	.05
Potash	.67	.61
Water and organic matter	3.12	4.10
Total	100.00	100.00

MINES AND MINERALS OF WASHINGTON.

PART FIFTH.

SECOND ANNUAL REPORT OF GEORGE A. BETHUNE, STATE GEOLOGIST.

that great territory known as Oregon until subdivided by the government, when two great and separate political divisions were created, namely, Oregon and Washington. At the remote period of this occurrence the "Okanogan country" was known as such, and was depicted by numerous writers of that day as "a vast wilderness infested with roaming bands of savage men, who gained a livelihood hunting over its bunch grass hills the game that there abounds and fishing its streams, and, latterly, bartering and trading the furs of the various species of amphibious animals that in them were to be found."

It was under the territorial *régime* that the "Okanogan country" first attracted attention, and the great resources of the district, both of a mineral and agricultural character, became known outside of a coterie of old timers, who, content in the luxuries of an existence incident to a residence within its boundaries, were neglectful of its competency to maintain in comfort not alone its full quota of inhabitants but, if necessary, a commonwealth as well.

These great resources of this great country becoming known, methods of ascertaining their permanency were speedily undertaken and it took but a short time to practically demonstrate that one of Washington's richest possessions was this "Okanogan country." The news heralded, attention was riveted on the country. As a result, in the year 1889, the "Okanogan country" that was, became the Okanogan county of to-day.

Okanogan county is the largest political subdivision of the State of Washington; and its metes and bounds may be described as follows: To the northward, by the British possessions; to the eastward, by the Colville Indian Reservation (executive order—July 2, 1872) and the Columbia river; to the south, by the rich mineral and agricultural counties of Douglas and Kittitas; to the westward by the counties of Whatcom, Skagit and Snohomish. In extent of area, Okanogan county exceeds Rhode Island and equals three other of the states in the union in size.

It is my province, as state geologist, simply to treat of matters pertaining to my department of this state's government. But I desire here to state that, while in my opinion Okanogan county is chiefly valuable for its mineral resources, the county will rank any other political subdivision of the state as regards the productiveness of its soil from the standpoint of the agriculturist. As a fruit growing region, it is second to none in the Pacific Northwest. As a cereal producer it stands unexcelled. Garden products, even those of a tropical zone, here thrive with the luxuriousness of the Floridian clime. Old Chief Moses's reservation, old home of the aborigine, goal of the trapper of early days, ranks to-day one of the most, if not the most, important counties of the state. Remote as yet it is from any of the principal, even the minor, commercial marts of Washington, even since the issuance of my last report it has taken rank as a prime, a most important, factor in the upbuilding of the state, and to it the state looks for that measure of sustenance, which, I am of the opinion, Okanogan cannot only yield in full measure, but leave in its garners an abundance for the proverbial "rainy day."

TOPOGRAPHICAL.

An idea of the topography of Okanogan county may possibly best be had by considering it in this regard from the Columbia river, one of the features of its eastern boundary line. From the Columbia, trending toward the northwest, its area takes a general rise, as it does also but to a lesser degree toward the Cascade range of mountains. Briefly, Okanogan county, in this regard, comprises a series of undulating plains, valleys and hills, these terminating on the county's northern, northwestern and western borders in a broken chain of mountains that it would seem were nature's boundaries of the Okanogan.

I shall now proceed to report upon, *seriatim*, the several mining divisions of this county, referring first to that of

RUBY MINING DISTRICT.

GEOLOGICAL FORMATIONS.

The geological aspect of this pioneer district is not at all peculiar, its principal bearing being in consonance with the formation of the general mineral bearing area of the Pacific Northwest mineral region. Its country rock may be said to be, in fact, identical, of course with some differences, but with none of marked character.

Here, for the most part, we find the country rock to be granite, gneiss, syenite, mica and hornblendic schists. Nearly vertical, uplifted and rolled together, these show a roughly bedded structure, striking from the northeast to the southwest, varying from 80 degrees to an approximation of the perpendicular. At very oblique angles of from 12 to 18 degrees, cutting through Ruby mountain's ridge, are these bedded lines of granitic schists. The longitudinal axis, as compared with the course of the bedded plains of the rock, is from 12 to 18 degrees more to the southward.

The appearance of the country rock in general, especially where denudation has occurred, shows a feldspathic granite of coarse texture, hornblendic and micaceous schists taking turn about and traversing the granite in bands. No true slates, so far, have been found in the district. Investigation made seems to indicate that the formation is of the primordial type, and is possibly traceable to the archean series.

THE MINERAL BELT.

The mineral zone, so far as is known, is mostly confined to the bedded structures described by me in the chapter referring to the geology of the district, and in width is from four to five miles, being about that distance in length, and encompassing an area approximating seventy square miles.

Several of the most prominent mines in the district are conformable to the dip and trend of the schistose rocks, their trend being nearly southeast and northwest, varying from 70 degrees to nearly perpendicular positions. Local variations, both as to trend and dip, are numerous, despite the fact as stated, that the local country rocks hold their course.

A course nearly north and south is followed by two or three ledges in the district. Below the horizontal plane these ledges dip toward the east at higher angles, variations ranging from 50 degrees to 80 degrees being marked. At angles through the bedded formations these ledges cut obliquely, here varying from 20 degrees to 60 degrees, a demonstration that these are true fissure veins.

This district is located in the southern part of Okanogan county, thirty miles north of the Columbia river, to the westward of the Okanogan river about fifteen miles, and is bounded on its western line by the Chloride mining district. Its northern boundary is the Salmon River mining district.

The district was formally organized in 1887, by Messrs. T. D. Fuller, R. Dilderback, John Kladiskey, P. McGeel, James Milliken and John Clunan.

The first locations filed after the organization were for claims on what is known as Peacock hill, in the northwest part of the district and about two miles northwest of Ruby City, the principal town of the district. The first location filed was for a claim called the "Butte," its locator being John Gober. This location was filed in June, 1886, prior to the formation of the district. The next earliest location was that of Birch Bill, Fred. Wendt, M. Gaspel, Joseph Janglow, and James Lufor, the War Eagle claim, located on the southern base of Peacock hill, in September, 1886. Then followed locations for the Leonora by J. C. Robinson and J. Crawford, the Peacock claim by John P. Carr and Joseph Janglow.

Ruby mountain, lying southeast of Peacock hill, was the scene of the next early locations. These were as follows: The Poorman, Fairview, and Ruby claims, located by John Clunan and James Milliken, made in October, 1886. In the same month the First and Second Thought mines were located by John Kladisky, R. Dilderback and P. McGeel. These might be styled pioneer locations in what is now one of the banner mining districts of the state.

THE FIRST THOUGHT MINE

May be put down as the principal mine in Ruby district. While in this district I believe there are just as valuable properties as this, even, possibly, more so, the fact remains that a sufficiency of development has been accorded this property to demonstrate its value beyond paradventure, and the same cannot be said of other properties in the district. In short, the idea desired to be conveyed is, that by reason of the First Thought mine having undergone more development than any other in the district, and the showing being superior, it must take first rank.

Development work comprises tunnels, drifts, cross-cuts, levels and winzes, the total number of feet representing upwards of 4,000. To more thoroughly illustrate the method of operation employed, I will say the mine is situated on the western slope of Ruby mountain, and is being worked by the tunnel system. The mine has given employment, on an average, to thirty men, these averaging in earnings per diem $3.50.

The vein is a ribbon quartz, carrying gray copper, silver, iron pyrites and zinc; and in thickness varies from fifteen to sixty feet, with a pay streak along the hanging wall of about eighteen inches. The trend of the vein is north and south, being almost vertical.

From assays taken across the vein, returns were $30 in silver. From the pay streak some samples of fahlerz, or gray copper ore, gave respectively 1,100, 1,330 and 1,926 ounces per ton. Shipments of first-class ore from this vein milled 380 ounces per ton, in silver. Owners of the First Thought mine comprise the First Thought Silver Mining Company, a Portland, Oregon, organization, who propose erecting a mint for the treatment of its own ores. Luther Wagner is the manager.

THE ARLINGTON MINE

With reference to the First Thought mine, lies south of the same, and is located on the western slope of Ruby mountain, at an altitude of 4,400 feet. This mine was located by John Oleson in May, 1887, and passed into the hands of the present operators shortly afterwards.

The width on the surface of the vein here is about six feet, with a pay streak two feet in width. The ore is a quartz, carrying gray copper, iron pyrites and ruby silver. The trend of the vein is northwest and southeast, and it is almost vertical.

The Arlington is one of the best equipped properties in the district, being provided with a hoisting plant, etc. On the surface are to be found a good boarding house, and ore house.

Assays of Arlington ore returned:

No. 1. Per ton, silver, $10; gold, trace.

No. 2. Ore from a distance of 120 feet from the location of the shaft in the southeast level, silver, $230 per ton; gold, trace.

No. 3. Ore from the end of the southeast level, silver, $219.30; gold, trace.

No. 4. Sample across vein at end of level, silver, $42; gold, trace.

No. 5. Sample from shaft, silver, $122; gold, trace.

No. 6. Samples from northwest drift ran, respectively, $113, $115, $111.30 and $1,500. The last named contained ruby silver.

No. 7. Average sample across vein, $40.

The Arlington Mining and Milling Company, a Portland corporation, owns this property.

THE FOURTH OF JULY.

This property is owned by the Fourth of July Mining Company, a Montana syndicate headed by Joseph Clark, esq., of Butte, and John Sheehan is in charge. Mr. Sheehan hails from Butte, Mont.

The original locators were B. Chilson, P. Pierce, P. McGeel and R. Dilderback, who discovered the mine in April, 1887.

The mine is located near the summit of Ruby mountain, at an altitude of 4.500 feet, and lies to the south of the Arlington mine.

The vein is of quartz, carrying brittle silver, ruby silver, horn and native silver. Its width on the surface is twenty-five feet. The mine has and is undergoing a thorough practical development, and as a result shows splendidly.

The workings comprise a double compartment shaft, down 200 feet, and a system of levels run on the vein. At the 100-foot level a drift has been run 200 feet south and 100 feet north. At the 200-foot level a drift has been run 100 feet south and 15 feet north. In what is known as the old workings of this mine a shaft is down 111 feet, with a level 100 feet north and 200 feet south. All the above named workings are in ore. In the old workings of the vein, where cross-cut, the vein matter is twenty feet wide, with a pay streak of four feet of ore. It was here that I found a portion of this pay streak exceedingly rich, there being eight inches of native silver ore. On the dump are about 250 tons of ore.

ASSAYS.

Assays of Fourth of July ore gave returns as follows:

No. 1. From the pay streak in old workings returned, silver, $870; gold, trace.

No. 2. From across the four-foot pay streak returned, silver, $413; gold, trace.

No. 3. From across the vein, silver, $172; gold, trace.

No. 4. In 100-foot level, new workings from south drift, silver, $482; gold, trace; from north drift, silver, $37; gold, trace.

No. 5. In 200-foot level, south drift, new workings, silver, $413; gold, trace; north drift, silver, $120; gold, trace.

The equipment comprises a hoisting plant, provided with a thirty-five horse power boiler and engine, ore house, boarding house, office, etc.

The number of men employed averages fifteen, the wages paid averaging $3.50 per diem.

The mine is shipping per month, on an average, ten tons of high grade ore.

OTHER DISTRICT PROPERTIES.

Having at some length proceeded to describe the principal mines in Ruby district, I shall now report upon those of lesser note, but which are very promising prospects.

The Leonora vein is situated on the southern slope of Peacock hill, and is looked upon as the principal claim on the First Thought vein. It is the property of Hon. J. T. McDonald and J. C. Robinson, of Ellensburgh. It was located by the present owners in September, 1886. The vein is five feet wide, with a pay streak of four inches, carrying brittle silver, galena, and iron pyrites. Workings consist of a shaft sunk on the vein to a depth of eighty feet. On the dump are about thirty tons of ore. Assays of ore from this claim were as follows:

No. 1. From surface, $150 silver; 20 per cent. lead.
No. 2. From along the vein, $138 silver; 25 per cent. lead.
No. 3. At 60 feet from surface, $350 silver; 17 per cent. lead.
No. 4. Bottom of shaft, $280 silver; 19 per cent. lead.

THE PEACOCK CLAIM.

The Peacock claim is located on the summit of McDonald peak, and is the northwest extension of the Leonora. The Peacock is the property of Thomas Hanscomb *et al.* It was located by John P. Carr and Joseph Janglow in October, 1886, prior to the organization of the district. The vein crops on the surface about three feet in width. The workings consist of a tunnel driven 170 feet for the purpose of cutting the vein. The ore is a quartz, carrying galena and iron pyrites. Assay returns from this ore were:

Minerals.	*Surface.*		*Dump.*	
Silver, ounces	34	40	112	Per ton, 2,000 lbs.
Gold, ounces	tr.	tr.	5	
Lead, per cent	16	16½	11	
Iron, per cent	10	10	0	

THE WASHINGTON.

The Washington is the northwestern extension of the Peacock. No authentic account of the date of location of this claim has been obtained. It is, however, owned by the same parties controlling the Peacock claim. The topography of the section in which this claim is located is peculiar. Peacock hill there abruptly terminates, and a sheer descent of 2,500 feet to the south fork of Salmon river is had. At this point a tunnel could be driven which would tap the Washington, Peacock, Leonora and several other veins at a depth of 2,500 feet from the apex of the Leonora vein. In width of vein and character of ore, those of the Washington are similar with the above properties.

THE WAR EAGLE.

The War Eagle claim lies west of the Leonora vein, of which, in fact, it is an extension. This claim was located in September, 1886, by Fred Wendt, M. Gaspel, Joseph Janglow, James Lee and Birch Bill. It is now the property of the War Eagle Gold and Silver Mining Company, a St. Paul corporation.

The width of the vein is nine feet, with a pay streak of two feet in width, carrying gold, silver and lead. The workings comprise a cross-cut near the south end line of the claim; two shafts of a depth of 100 feet each through ore; at the bottom of these a 200-foot level connecting them, and the stripping of the vein on the surface. There are on the dump 500 tons of ore. Assay returns were as follows:

Minerals.	*Surface.*				*Bottom Shaft.*			*Dump.*	
Silver, ounces	40	62	31	91	29	32	67	65	Per ton, 2,000 lbs.
Gold, ounces	10	tr.	tr.	2	tr.	8	6	tr.	
Lead, per cent	10	...	tr.	2	4½	9.12	6.30	1½	

This mine is rendered easily accessible from the main county road by private roadway.

THE IDAHO CLAIM.

The Idaho claim is the easterly extension of the War Eagle, above described. This claim belongs to the Idaho Mining Company, a Spokane corporation, and has been in charge of George E. Pfunder, a Colorado miner, now mineral collector for the Washington World's Fair Commission. It was located, I believe, about the autumn of 1886, shortly afterward passing into the hands of its present owners.

The width of the vein here is fourteen feet between walls, with a pay streak of two feet. The character of the ore is a quartz, carrying gold, silver and lead. Thus far a shaft fifty feet in depth and a cross-cut at the bottom thereof, exposing both walls, comprises the workings on the property. Assays, taken promiscuously from the vein where exposed, returned:

Minerals.	Surface.			Bottom Shaft.		
Silver, ounces	42	47	54	34	39	35.4
Gold, ounces	tr.	tr.	tr.	tr.	tr.	tr.
Lead, per cent	6	9	4	7	7½	3¾

THE POORMAN CLAIM.

The Poorman claim is the western extension of the Idaho claim. The date of the location of this claim was October, 1886. The locators were those veteran prospectors John Clunan and James Milligan. It is now owned by the former, W. H. Singleton, Thomas Donan and Hon. H. W. Fairweather.

The width of the vein is four feet, with a pay streak of about three inches. The character of the ore is a quartz, carrying galena, iron pyrites, gold and silver. The strike of the vein is north and south, with a dip to the east.

The workings consist of one tunnel 100 feet in length, tapping the vein at a depth of 150 feet. Assays made from samples of this ore returned:

Minerals.	Surface.			Dump.			
Gold, ounces	tr.	tr.	tr.	tr.	tr.	tr.	Per ton, 2,000 lbs.
Silver, ounces	38	36	34	34	31	40	
Lead, per cent	25	33	21	18	24	21	

THE FAIRVIEW PROSPECT.

The northern extension of the Poorman, a claim owned also by the parties named, has been called the "Fairview." The Fairview ore is similar to that of the Poorman, and in size of vein and pay streak, and in assay returns the two are similar.

THE SHELBY CLAIM.

The Shelby claim is one located on the west side of Peacock hill. The owners are J. P. Fogarty, Edward Pryun and William Stedman.

The vein is six feet in width, and is a quartz, carrying galena and arsenical iron. This is a milling ore. The work completed consists of an incline shaft driven a distance of seventy-five feet. Assays of ore returned:

Minerals.	Shaft.			Dump.			
Silver, ounces	140	38	41	120	154	29	Per ton. 2,000 lbs.
Gold, ounces	tr.	tr.	tr.	tr.	tr.	tr.	
Lead, per cent	4	4½	.5	5	4½	.6	
Iron, per cent	......	......	7	......	......	7.5	

On the dump are about twenty tons of ore.

SILVER STAR PROSPECT.

The Silver Star claim, the southern extension of the Shelby, is owned by B. Dougherty and Wm. Stedman. The vein on this claim is four feet in width. The ore is of the same character as found in the Shelby. The strike of this vein is north and south, dipping 60 degrees to the east. An incline shaft has been driven a distance of seventy-five feet through ore. There are about twenty tons of ore on the dump.

OTHER PROPERTIES UNDERGOING DEVELOPMENT.

To describe each and every one of "the hundred and one" prospects in this district would fill volumes. On each a certain measure of development is to be noted. Some, by reason of a greater amount of work completed upon them, look better than do others less fortunate in that regard. However, a universality of opinion will be found existent among their owners, that all are mines in embryo, and the fact that lapsed locations are few and far between in the district, and that assessment work is hardly ever neglected, speaks volumes as regards the confidence owners of prospects, likely looking and otherwise, have in their possession. While, naturally, some are good, others indifferent, and others, in the opinion of mining men, practically valueless, there can be no doubt that the confidence, hope and trust of the several possessors of all are in their properties. But there are many prospects, using that term for just what it implies to the miner, and no more, in Ruby mining district that, to my eye, look as well in their embryotic state as do some of the full-fledged mines of the district, and among these are the following:

The Buckeye claim, owned by George Melvin, on Ruby hill, a seven-foot vein of quartz ore, samples of which assayed $100 to the ton in gold and silver.

The Arizona, owned by the same party, and an extension of the Buckeye, with a vein six feet wide of a like character of ore, assaying also $100 per ton. One hundred and forty feet of work has been done on these two prospects by Mr. Melvin.

The prospects of the Keystone Mining Company, a Portland corporation, look especially well, and properly handled should develop into valuable properties. On the Keystone, one of its claims, a shaft 150 feet deep has been sunk. Assays of ore taken from this shaft returned in gold and silver $45 per ton.

The Anaconda and Bonanza King also deserve mention as likely looking prospects; and also the properties of the Ruby Hill Mining Company. Aside from the above named, are the following prospects which may or may not turn out to be valuable properties:

The Northern Light, The Copper Queen, El Dorado, Modoc Chief, Denver, Second Thought, Wheel of Fortune, Original, Pomeroy, and The Missing Link.

SALMON RIVER MINING DISTRICT.

But a year marks the difference, as regards their ages, between Ruby mining district and one six miles to the southward in Okanogan county, known as the Salmon River mining district, christened by its founders after the fine stream that flows through its area. Salmon River district ranks Ruby in age by a twelvemonth.

A feature of this district is its accessibility, there having been provided for the miner five natural causeways throughout its domain. Of the country comprising Salmon River district it may be said it is the most open of any in Okanogan county. The prospector may enter within its gates and traverse it from either principal point of the compass with comparative ease. There are no high mountains to obstruct his path or hinder him in making a rapid research of its mineral resources. Then, too, water, various species of game and fuel in abundance are to be found on every hand. Salmon River district is a veritable paradise for the prospector in so far as comforts—even luxuries—to be gleaned from the surface of its area are concerned. But, most important of all, it is not behind other mineral subdivisions of the county in the wealth of precious metals to be found within its confines. The district ranks one of the richest in the county in this respect.

HISTORY OF ORGANIZATION.

George W. Forrester, J. C. Boone, P. Pierce and John Gober founded Salmon River mining district in the year 1886, shortly after that veteran prospector, Mr. George Runnels, heralded the news that he had made rich discoveries of mineral within its present borders. Since that time the district has steadily risen to a seat of prominence as a mining section, and to-day its prospects as a coming leading producer of gold and silver are of a most flattering character. Many of the best known mines in the Northwest are to be found there, and not a month passes but what the news of valuable discoveries is heralded. The population of the district has increased enormously the past year, and next year will witness a veritable hegira to that section. The open character of the country, the advantages of the district in the shape of accessibility, abundance of fuel and water, and, most important of all, the demonstrated richness of the mineral deposits there abounding, are factors that have tended to and will continue to draw within the district the best classes alike of investment seekers, operators, miners and prospectors.

SALMON RIVER MINERAL.

The country rock of this district is mainly granite, gneiss, syenite and micaceous schists. The principal mines of the district lie upon both sides of the Salmon river, on what are known as Mineral hill and Homestake mountain; the former being on the east side of the river, the latter on its western side. Mineral hill is a peak rising from Salmon river to an altitude of about 6,000 feet above the level, and about 4,000 feet above the bed of Salmon river.

Homestake mountain, so called after a mine of that name located upon it, is about 6,200 feet in height, its altitude above the river bed being 4,200 feet. Like Mineral hill, Homestake mountain's contour might be described as a succession of benches, one upon the other from base to summit.

Among the many promising properties on Mineral hill is what is known as the John Arthur mine.

THE JOHN ARTHUR MINE.

This property, one of the coming claims of the district in my opinion, is situated on the east side of Mineral hill at a point about a mile north of the county seat of Okanogan county. The John Arthur was the location originally of Mr. J. C. Robertson of Conconully, and was the pioneer location on the hill, being made in the month of June, 1886. Hon. J. T. McDonald, of Ellensburgh, is here largely interested.

The Arthur vein is ten feet in width, with a very large outcrop on the surface. The pay streak will average in width two and one-half feet. The strike of the vein is northeast and southwest, and the dip about 60 degrees to the west.

The mine is being worked by an inclined shaft which has been driven to a depth of about seventy feet in ore.

On the dump were about fifty tons of ore, which in character is a quartz, carrying iron pyrites, wire silver and silver glance. Assays of ore from the mine returned as follows:

Minerals.	*Surface.*			*Shaft.*			*Dump.*			
Silver, ounces	30	35	82	101	98	98½	80	94	116	Per ton, 2,000 lbs.
Gold, dollars	2	tr.	5	tr.	tr.	4	2	tr.	2½	

The systematic development of the property has had much to do with the excellent showing it has made, and, if continued, should be attended with rich results.

THE LONE STAR MINE

Is the property of the Lone Star Mining Company, a Tacoma corporation, and is an extension of the John Arthur mine just described.

The width of the vein averages three feet between walls. The pay streak is about one foot in width. The ore is a white ribbon quartz, containing galena, chalcopyrite (sulphide of copper) and brittle silver. On what is known as the main workings, considerable stoping has been done, and a quantity has been piled upon the dump ready for shipment. These main workings comprise an inclined shaft about 375 feet in depth, and a system of levels, driven north and south along the vein. Surface equipment for the main workings is represented by a hoisting plant, shaft and ore houses.

The old or upper workings of the mine comprise about 450 feet, in the shape of tunnels, cross-cuts and levels. These workings were driven for prospect purposes. Assay returns of samples of ore from the croppings, seventy-five foot level, eighty-six foot level, sixty-five foot level, seventy-eight foot level and twenty-five foot level, were as follows:

Level.	Gold....	Silver, ounces....	Lead, per cent......
Croppings, $115 per ton.			
75-foot level (north), 100 feet deep	.1	63	32.00
86-foot level (south), 100 feet deep	.1	65.8	44.16
65-foot level (north), 200 feet deep	.1	70	39.00
178-foot level (south), 200 feet deep	.3	40	52.00
25-foot level (south), 300 feet deep	tr.	140	65.00
Bottom dump, 375 feet deep	1.6	39.6	30.00
Approximate average	.3	65	40.00

Average worth per ton, 2,000 pounds $80.00

THE TOUGH NUT MINE.

This mine is located on the western slope of Homestake mountain, in what is known as Tough Nut Gulch, on the eastern side of Salmon river.

The location of the property dates back to May 10, 1886, when Messrs. Forrester, Pierce, Gober and Boone, original discoverers of the wealth of the district, filed upon it. The mine is at present the property of Messrs. Charles Uhlman of Tacoma, William B. Kelley of Sumner, W. J. Thompson of Tacoma, and George W. Forrester of Conconully, Okanogan county.

The Tough Nut vein represents an immense "blow out" of ore, at the surface measuring twenty feet between walls. The workings consist of an inclined shaft driven through ore a distance of fifty feet. At a distance of 100 feet below the mouth of this inclined shaft, a cross-cut has been run in 220 feet, tapping the vein. Here the width of the vein measures five feet, with a pay streak of three feet, the latter being solid galena. On the dump there are about 200 tons of ore.

Assay returns from surface, inclined shaft, cross-cut and dump samples returned as follows:

Minerals.	*Shaft.*	*Cross-cut (breast).*	*Dump.*	
Croppings, $50 per ton.				
Gold, ounces	....	....	....	Per ton, 2,000 lbs.
Silver, ounces	47	86.00	65	
Lead, per cent	32	43.16	41	

THE HOMESTAKE CLAIM.

This property, after which Homestake mountain takes its name, was located at about the same time, and by the same parties. The vein here is, at the croppings, eleven feet in width. A tunnel has been driven 175 feet along the vein. A cross-cut forty-two feet in length, and a shaft in depth twenty-nine feet, comprise the development work thus far completed on this property. The ore is a quartz and galena, carrying gold and silver. About 300 tons of ore are on the dump.

Assays from croppings, shaft, tunnel and dump returned:

Minerals.	*Shaft.*	*Tunnel.*	*Dump.*	*Croppings.*	
Gold, ounces	2	tr.	....	tr.	Per ton. 2,000 lbs.
Silver, ounces	18	54	48	56	
Lead, per cent	....	32	36	36	

The Homestake is the property of I. W. Anderson, B. R. Everett and Otis Sprague, all of Tacoma, who are applying for a patent for the property.

THE SALMON RIVER GROUP

Of mines comprises the Salmon River Chief, the Wellington, the Knickerbocker, the Salmon Creek and Manhattan claims, and they are now all controlled by Mr. Henry Wellington of Okanogan county, although previously they were owned by Mr. Wellington and Messrs. W. Daniels, J. C. Boone and Thomas O'Neill. The group is located less than a mile from Conconully, and its members form a cluster of the most promising claims in the district.

On the Salmon River Chief is a vein three feet in width of quartz, carrying galena and silver glance, with a pay streak twelve inches in width. Development work consists of a tunnel driven a distance of fifty feet on the vein.

Assays taken from the croppings, tunnel and dump, returned:

Minerals.	*Croppings.*	*Tunnel.*	*Dump.*	
Gold, ounces	.10	trace	.10	Per ton, 2,000 lbs.
Silver, ounces	83.00	72	75.00	
Lead, per cent	12.00	13	10.00	

The Wellington, second claim of the group, is located 200 feet above the Salmon River Chief, at an altitude of 2,800 feet above sea level. Here the vein is shown twelve inches thick at the croppings, and is a quartz, carrying galena, silver and gold. The workings consist of one tunnel driven a distance of thirty feet on the vein. In the breast of the tunnel the ore opened up to four feet in width. On the dump about five tons of ore were found.

Assays of samples from croppings, tunnel and dump, returned:

Minerals.	*Croppings.*	*Tunnel.*	*Dump.*
Gold	.10	4	2
Silver	58.00	63	60
Lead	12.00	14	14

The Knickerbocker, third member of the group, is the northwest extension of the Wellington, and the ore is similar in character. The croppings here show a vein three feet in width. There has been nothing but assessment work completed thus far on this property. Assays of samples from the croppings returned, per ton (2000 pounds), silver, 50 ounces; lead, 10 per cent.

The Salmon Creek, the fourth member of the group, is located above the Knickerbocker, and at an altitude of 6,000 feet above sea level. The width of the vein is about five feet, and the pay streak about 18 inches. Workings consist of a tunnel run in twenty feet on the vein, and a shaft has been sunk a distance of seventy-five feet in ore. The ore is a galena, gray copper and arsenical iron. There are about ten tons of ore on the dump. Assay returns from tunnel, shaft and dump, returned:

Minerals.	*Tunnel.*	*Shaft.*	*Dump.*	
Gold, ounces	tr.	tr.	41	Per ton, 2,000 lbs.
Silver, ounces	43	41	63	
Lead, per cent	18	18	20	
Iron, per cent	6	7	7	

This year (1891) a tunnel was started below the Wellington claim a distance of 300 feet, for the purpose of tapping the vein of this claim, and also that of the Salmon Creek.

The Manhattan lode, last of the group, lies northwest of the Salmon Creek claim. Here the vein is four feet in width, and is a gold, silver and lead proposition. The work completed consists of a tunnel run in on the vein a distance of about forty feet. There were about forty tons of ore on the dump. Assay returns on samples from tunnel and dump returned:

Minerals.	*Tunnel.*	*Dump.*
Gold, ounces	tr.	tr.
Silver, ounces	38	36
Lead, per cent	6	6½

THE LADY OF THE LAKE

Is a mine situated on the southern extremity of Homestake mountain, north of Conconully lake, and is one of the oldest claims in the district, having been located May 10, 1886, by Mr. George Runnels, the pioneer prospector of the district. The mine is now the property of Messrs. Hardenburgh and Higley, of Conconully. The vein of the Lady of the Lake is a decomposed quartz, carrying gold, silver and lead, and the croppings show eighteen feet of this ore. The work completed this year comprises an inclined tunnel driven a distance of eighteen feet on the vein, in the upper workings. At a distance of 350 feet below this, and down the hill, a tunnel has been driven in on the vein a distance of fifty feet, tapping a rich pay streak of four feet of rich quartz, carrying galena and brittle silver. There are about 350 tons of ore on the dump. Assays from samples from upper workings, lower workings and dump were as follows:

Minerals.	*Upper workings.*	*Lower workings.*	*Dump.*	
Gold, ounces	tr.	tr.	tr.	Per ton, 2,000 lbs.
Silver, ounces	70	125	90	
Lead, per cent	15	11	13	

Properly developed, the Lady of the Lake should rank a leading producer in the district.

THE MINNEHAHA,

Of Salmon River mining district, lies on the west side of Homestake Mountain, about 1,500 feet above Salmon river, and is an extension of the Tough Nut claim. The vein on the surface is seventeen feet thick between walls, and is a quartz, carrying galena, iron pyrites and gray copper. There are several rich streaks of ore running through the vein. The workings comprise a cross-cut 200 feet in length, tapping the vein at a depth of 100 feet. A tunnel could be driven 600 feet below the present workings, which would tap the vein at a depth of over 700 feet. Assays of samples obtained from both cross-cut and dump returned:

Minerals.	*Cross-cut.*	*Dump.*	
Gold, ounces	tr.	tr.	Per ton. 2,000 lbs.
Silver, ounces	113	98	
Lead, per cent	14	12½	

The Minnehaha belongs to George Cooper and T. A. Wilson, of Conconully.

THE EVENING STAR

Claim is the northern extension of the Minnehaha, just described, and is owned by Messrs. Segmar and Manual, of Conconully. Here a vein four feet in width is shown on the surface. The vein is of native silver and copper ore. The workings comprise a shaft of sixty feet on the vein.

THE OKANOGAN BELLE

Is the northern extension of the Evening Star, and is owned by F. S. Hines, of Tacoma. The vein is thirteen feet in width. The character of the ore is a quartz, carrying galena, iron sulphide and silver glance. Workings comprise a cross-cut 150 feet in length, tapping the vein. There are about thirty tons of ore on the dump. Assays from samples from cross-cut and dump, returned as follows:

Minerals.	*Cross-cut.*	*Dump.*
Silver, ounces	92	95
Lead, per cent	10	10

I have now described, aside from several fine looking prospects, all the principal properties located on Homestake mountain, and shall proceed next to report upon the several leading mines on Mineral hill. Referring first to

THE MONITOR CLAIM,

A property located at an altitude of 5,500 feet on the hill on its southern flank. The owner is Mr. Robert Hargrove, of Conconully. The Monitor vein shows one foot of quartz and galena on the surface, the vein widening out, at a depth of fifty feet, to three feet. On the surface the vein has been stripped a distance of several hundred feet. Workings comprise a shaft sunk a depth of fifty feet through ore. On the dump are about ten tons of the ore described. Assay returns from shaft and dump were:

Minerals.	*Shaft.*	*Dump.*	
Silver, ounces	30	35	Per ton, 2,000 lbs.
Lead, per cent	18	17	

THE MAMMOTH CLAIM

Lies northeast of the Monitor, and is the property also of Mr. Hargrove. Here are to be found two distinct veins. The upper vein is five and the lower six feet in thickness. The character of the former is a quartz, carrying gold, silver and copper; of the lower, a quartz, carrying gold and silver. On the upper vein, a cross-cut thirty feet in length has been completed, as has a like cross-cut on the lower vein. From the first named Mr. Hargrove made the first shipment of ore ever sent to the smelter at Helena, Montana, for treatment from Salmon river mining district, an event occurring in the month of December, 1889. The ore netted Mr.

Hargrove $250 per ton. Assays of samples taken from both cross-cuts on both veins follow:

Minerals.	*Upper vein.*	*Lower vein.*	
Gold, ounces	3.10	1.10	Per ton, 2,000 lbs.
Silver, ounces	330.00	87.00	
Copper, per cent	7.13	....	

On the dump of this claim are about fifteen tons of ore.

THE LAKEVIEW,

Another of Mr. Hargrove's claims, lies northeast of the one just described, and has a vein showing two feet at the croppings, of quartz and sulphuretts. A tunnel 130 feet in length has been driven on a "stringer" of the main vein.

The main vein of these claims, which are known as the Hargrove group of mines, could be reached by a tunnel driven a distance of 500 feet, which would tap the vein at a depth of 1,000 feet.

THE HARDSCRABBLE

Claim parallels the Lakeview claim, and is owned by the Hardscrabble Mining and Milling Company. The vein crops a width of about fifteen inches and has been stripped in several places along its course. Work comprises a tunnel thirty feet in length, on the vein. There are several tons of ore on the dump. Assay returns of ore were:

Minerals.	*Shaft.*	*Tunnel.*	*Dump.*	
Gold	....	....	....	Per ton, 2,000 lbs.
Silver, ounces	178	335	658	

THE MOHAWK CLAIM

Is the property of Henry Lawrence and others, and is located on the southern side of Mineral Hill. The ore is a quartz, carrying galena and copper sulphide, and the vein will average about three feet in width. This vein has been stripped on the surface in several places. The workings consist of a tunnel driven in 200 feet on the vein. On the dump are about ninety tons of ore. Assay returns from breast of tunnel and dump returned:

Minerals.	Breast of Tunnel.	Dump.	
Gold, ounces	tr.	tr.	Per ton, 2,000 lbs.
Silver, ounces	62.00	65.00	
Lead, per cent	7.00	5.00	
Copper, per cent	3.10	3.12	
Iron, per cent	....	8.70	

THE EUREKA CLAIM

Is located on the western slope of Mineral hill, and is the southeasterly extension of the Lone Star vein. The claim is that of Richard Malone and George Gubser, of Conconully. There is a four-foot vein of quartz on this property, carrying galena, copper and iron pyrites. A shaft has been sunk on the vein, in ore, to the depth of sixty-five feet. Assay returns from samples of ore taken from this shaft returned 80 ounces silver and 13 per cent. lead per ton (2,000 pounds).

THE WASHINGTON CLAIM

Is the southern extension of the Eureka claim. Here a vein on the surface measuring seventeen and one-half feet has been found. The ore is of similar character to that found in the Eureka, and assays about the same. The work done comprises a shaft down sixty-five feet, and further down the hill a tunnel is being driven to tap the vein. Richard Malone and George Gubser also own this claim.

THE LAUNA CLAIM

Belongs to Chester F. Griesmer *et al.*, of Tacoma, and was located July, 1886, by Lemon, Hunt and Stonsburg, and is situated near the summit of Mineral hill. The vein at the surface is about seven feet in width and is a quartz, carrying brittle silver, gray copper and silver glance. Improvements consist of two shafts — one down fifty feet, the other seventy-five feet — on the vein, in ore. A tunnel sixty feet in length has also been driven on the vein, in ore.

On the dump are about fifty tons of a high grade ore. Assays of samples from the bottom of each shaft, tunnel and dump returned:

Minerals.	*Shaft No. 1.*	*Shaft No. 2.*	*Tunnel.*	*Dump.*	
Gold, ounces	tr.	tr.	....	tr.	Per ton. 2,000 lbs.
Silver, ounces	860.00	812.00	130.00	200.00	
Copper, per cent	1.13	2.12	....	2.12	

THE SUNRISE CLAIM

Is owned by J. C. Collins and Newton Campbell, and is a quartz vein, carrying brittle silver, gray copper and silver glance ore. The vein is four feet between walls, with a pay streak of fourteen inches. A tunnel has been driven on the vein a distance of seventy-five feet in ore. The altitude of the property is 5,200 feet above sea level. Average assays of ore from the pay streak yielded 379 ounces in silver; from across the vein, 120 ounces of silver, per ton. There are about ten tons of ore on the dump.

THE GROVER CLEVELAND

Is the title given a claim located on Mineral hill. The Cleveland has a vein of galena ore four feet in thickness. On this claim two shafts have been sunk, one to a depth of thirty, the other to fifty feet. Average assays of ore from the bottom of these shafts returned fifty ounces silver, sixty per cent. lead, and five per cent. copper, per ton (2,000 pounds).

THE BRUNSWICK CLAIM.

Like the Cleveland, is located on Mineral hill. Here a twenty-inch vein of quartz and galena ore is found. A tunnel 120 feet in length has been driven on the vein. Average assays of this ore returned 225 ounces silver, and forty-five per cent. lead, per ton (2,000 pounds).

Other prospects, mines in embryo I might style them, but all good-looking beginners, are: The Blue Daisy, Thompson, Diamond, Hidden Treasure, Standard, Mayflower, Little Giant, Granite, Ajax, Telephone, Daisy, Poorman, Ida May, Bonanza, Banner, Mug, Glory, Mount Bay, Pacific, Buckhorn, Alta, Montana, Snowflake, Pointer, Onward and Boomer claims.

WANICUTT LAKE DISTRICT.

In my first annual report, "Mines and Minerals of Washington, 1891," I went into the history of this mineral subdivision of Okanogan county at length. In this report I shall simply outline its boundaries, and describe the measure of development accorded mining claims within them; also, refer to new discoveries, which, during 1891, have not been few and far between.

Wanicutt Lake district takes its name from a lake of that name situated within its confines. It is bounded on the north by Mt. Chapaka district; east by the Okanogan river; south by Galena mining district; west by the Sinlahekin river. It was founded about the year 1888 its formal organization as a mining district being directly the result of important mineral discoveries made in its area, in 1887 on Palmer mountain, scene to-day of the greatest measure of development of mines in the district. To Messrs. Robert Dixon and John Hunter is due the distinction of having made the first location under the laws of Wanicutt Lake district. These gentlemen, in June, 1887, discovered and located the Black Bear mine on Palmer mountain. Since then hundreds of locations have been filed for discoveries in this district. With the increased attention being paid to mines and mining in that region, I look for a decided revival of interest in Wanicutt Lake district, as there is a vast amount of mineral stored away there.

"THE GOLD QUARTZ BELT."

As I stated in my previous report, Wanicutt Lake district might be aptly called the "gold quartz belt of Northwestern Washington," but while the majority of properties there are gold quartz propositions, some very promising silver mines have been opened. One of these is a mine, which by reason of its richness and the magnitude of its mineral deposits, has become famous throughout the country. I refer to the Ivanhoe mine, also on Palmer mountain. The district has the past year engrossed more attention from capitalists and mining men than has ever before been accorded it. That "outsiders" are believers in the richness of its mines is evidenced by the fact that good prices have been offered for several properties there. That owners rest confident concerning the value of their properties is also evidenced by the fact that few, if any, Wanicutt Lake mines are on the market.

ACCESSIBILITY.

Wanicutt Lake district may be easily reached from Conconully by a series of good wagon roads, the district being but thirty miles distant from the county seat. It may also be reached from Coulee City, terminus of the Washington Central Railroad, over a good road. From Coulee City it is distant about 125 miles.

WATER AND WATER POWER.

The district is well supplied with water, the Sinlahekin river traversing it. Numerous mountain streams are also to be found. Along the summit of the mountain ranges bordering it is to be found an abundance of timber suitable for both mining and architectural purposes.

COUNTRY ROCK.

The country rock of the district is syenite, hornblendic schists, highly metamorphosed slates, quartzite and dolomitic lime.

PRINCIPAL MINES.

As stated, the principal mines of the district are located on Palmer mountain. This mountain is almost in the center of the district, rising abruptly from the Sinlahekin river to an altitude of about 4,500 feet. Trails and wagon roads are methods used in reaching the several mines upon it, the mountain, despite its appearance of abruptness, being found easily accessible on close approach. The principal mine on Palmer mountain is

THE IVANHOE MINE.

This now famous mine was discovered in the month of June, 1888, by Messrs. Al. C. Cowherd and Frederick Western, the present owners of the property.

The Ivanhoe is essentially a silver and gold proposition, and has yielded more bullion than any other mine in Okanogan county. The ledge crops for a distance of 1,200 feet, and for that distance is two and one-half feet wide, with a pay streak of two and one-half inches. The ore is a white quartz, carrying ruby, brittle and horn silver. The strike of the vein is northwest and southeast, dipping westward into the mountain at an angle of sixty degrees. The hanging wall is of slate, the foot wall hornblendic schists.

The workings of the Ivanhoe consist of two inclines running through the vein, in ore, sixty-five and eighty feet respectively. At the bottom of the latter incline the vein measures four and one-half feet between walls, the pay streak measuring two and one-half feet. At the bottom of the former incline the vein measures three and one-half feet; the pay streak two and a half feet.

On the surface a space 81 x 100 feet of the hanging wall had been denuded. The ledge at this point averaged three feet in width, and I approximate the quantity of ore there in sight at 2,000 tons.

One hundred feet from the above described point a shaft has been sunk to a depth of seventy feet, cutting the ledge and exposing to view a body of ore six and one-half feet in width. In lieu of steam power, a whim is employed in the operation of the mine.

The mine is provided with whim house, ore house, blacksmith shop and a comfortable boarding house. In and about it ten men are given steady employment, at wages averaging $3.50 per diem per man.

Regular shipments of Ivanhoe ore are being made to smelters in Washington and Montana, results of these being uniformly large. One shipment sent a smelter in Denver, Col., netted $636 per ton. Two shipments made the Tacoma smelter yielded $272 and $375 per ton, after freight charges of $45 per ton had been paid. On the dump are about 250 tons of ore.

BLACK BEAR AND WAR EAGLE.

These gold mines, which during the past year have attracted a deal of attention from mining men both at home and abroad, are found on the eastern slope of Palmer mountain, but a few miles distant from the town of Loomiston, in the district. Both are easily accessible by good wagon road, and both, until recently, were the property of a Spokane syndicate, headed by S. I. Silverman, who officiated as general manager at the mines. In January last these claims became the property of the Everett Mining and Milling Company, a corporation headed by F. W. Dunn, L. W. Getchell, Henry Hewett, jr., and others, the transfer to them having been effected by Col. James F. Wardner, Washington's leading mining promoter.

Both veins are identical as regards strike and dip. The strike is east and west, and the dip is nearly vertical. The ore carried is a gold quartz, carrying but a small per cent. of iron sulphide. On the property a five-stamp mill has been operated continuously for many months with profitable results. Assays of ore have averaged $50 in gold per ton of 2,000 pounds.

THE GOLD CROWN CLAIM.

On the wagon road east of the War Eagle mine is located the Gold Crown claim, formerly the property of Messrs. Lee, Jones and Turner, now owned by S. I. Silverman, of Helena, Montana. It is a gold quartz proposition. The Gold Crown vein is ten feet in width, with a strike northwest and southeast, and a dip of 50 degrees east, more or less. The main shaft is down thirty-four feet in ore. On the dump are about fifty tons of ore, assays from which returned $105 in gold per ton. Ore from this claim is handled at the five-stamp mill on the War Eagle-Black Bear.

THOMAS RYAN

Is the owner of the Bunker Hill mine, which at one time bid fair to rival as a gold producer any claim in the Northwest. It may yet rank as a great property if properly developed. The claim is located on the western slope of Palmer mountain, and was discovered by Messrs. Ryan and Nelson in 1890. The mine is situated at an altitude of 3,500 feet above sea level. Assays of the ore have returned as high as $300 to the ton.

THE KODAK AND EMPRESS

Are two claims on the northwest side of Palmer mountain, which are owned by F. W. Brown, J. Finn and J. E. Waters.

The Kodak vein trends northwest and southeast, with a dip of sixty degrees to the west, and is forty-five feet in width at the croppings. The character of the ore is a white quartz and sulphuretts, carrying gold and silver. The hanging wall is gneiss, and the foot wall mica schist. The work completed on the property comprises a tunnel driven in on the vein a distance of twenty feet. Assays of samples taken from croppings and dump were:

Minerals.	*Cropping.*	*Dump.*	
Gold, ounces	10	11	Per ton, 2,000 lbs.
Silver, ounces	80	90	

The Empress vein is about the same width as that of the Kodak, of which the former is the northwest extension. Work completed comprises the necessary assessment work required by law.

ON THE EASTERN SLOPE

Of Palmer mountain are five claims, known as the Rainbow Group, taking the name from the first of the claims located by the owners, Messrs. Morris Farrell, George Dorr and George McLaughlin, old time prospectors. This group has lately been bonded by the Washington Mining and Development Company, of Seattle.

The Rainbow, the principal member of the group, has a vein four feet in width, with a pay streak of about twenty-two inches. It is a contact vein, lying between a lime hanging wall and a quartzite foot wall. The trend of the vein is northwest and southeast, with a dip toward the west at an angle of twelve degrees. The ore is a white quartz, carrying copper and iron sulphides. The pay streak is rich in free gold. The work completed consists of a tunnel thirty feet in length on the vein, and a shaft forty-eight feet in depth through ore. There are about seventy-five tons of this ore on the dump. Assay returns of samples from vein matter and bottom of shaft returned:

Minerals.	*Vein Matter.*	*Shaft.*	
Gold, ounces	50	60.00	Per ton, 2,000 lbs.
Silver, ounces	20	18.00	
Copper, per cent	2	2.00	
Lead, per cent	3½	2.90	

Some exceedingly rich returns have been made by assay from the pay streak.

The Mayflower is the southern extension of the Rainbow. Work completed here represents the amount required as assessment work by law.

The Coyote is a parallel vein with the Rainbow. Here a three-foot vein of quartz, carrying gold and silver ore, has been found. Two cross-cuts on the vein represent the work completed, and there are about forty tons of free gold ore on the dump. Sample assays of the dump returned: No. 1, $120 free gold; No. 2, $35 free gold, per ton of 2,000 pounds.

The Cottonwood is the fourth member of the group. The vein

is two feet in thickness here, carrying free gold ore. An incline on the vein, through ore, is in a distance of thirty feet. On the dump are about forty tons of free gold ore. Assays of samples off this dump returned $32 in gold per ton.

The June Bug is the Cottonwood's south extension. It has a vein two feet wide of free gold ore. A shaft thirty feet in depth has been sunk, and there are about twenty tons of ore on the dump. Assays of samples of this ore went $69 per ton.

Properly handled, the above named five claims, comprising the Rainbow group, should make a great record as free gold ore producers.

THE ANACONDA MINE.

Close beside the shores of pretty Lake Wanicutt, in fact but 200 yards distant from its waters, is located a mining claim called the Anaconda by its discoverers in honor of the union's great Montana copper producer. However, the Washington namesake is a gold and not a copper producer. The Anaconda is the property of Captain Hall and George Pascall, gentlemen residing at Loomiston, a thriving new town of the district. It might more properly be described as being located on the eastern slope of Palmer mountain. The vein is four feet in width, with a strike northwest and southeast, and is composed of quartz, carrying iron pyrites.

Messrs. Hall and Pascall have run an incline a distance of forty feet in on the vein. Assays of samples taken from the dump returned $30 gold and $8 silver per ton of 2,000 pounds.

The western extension of the Anaconda is the Empire, also owned by Messrs. Hall and Pascall. A ten-foot shaft comprises the work here completed, and there are about five tons of ore of good quality on the dump, assay samples returning $18 gold per ton. The eastern extension of the Anaconda is also owned by Messrs. Hall and Pascall, they calling the location the Minnie. The vein of the Minnie is two feet in width. A shaft is being sunk on the vein. On the dump are about ten tons of ore. Assay, $22 gold per ton.

West of these properties is the Combination claim, of which Capt. Hall is sole owner. A four-foot vein of gold ore was here discovered. Capt. Hall has completed a shaft twenty-five feet deep on the claim, and about twelve tons of ore are on the dump. This ore on assay returned $21.50 per ton gold and $3 in silver.

THE IRON CAP CLAIM

Is the title given the discovery of a vein of quartz and hematite of iron ore by Oscar Nelson, its owner. This vein is seven feet in width, the ore carrying gold and silver. The strike of the vein is northeast and southwest, with a dip of forty-five degrees to the west. Mr. Nelson has completed an inclined shaft a distance of fifty feet on the vein. The hanging wall is slate, the foot wall lime. On the dump are about 100 tons of gold and silver ore. Assays of which returned $20 per ton in gold and silver.

The Cornucopia, under the same ownership, has a large vein of quartz ore, developed by a shaft sunk at a depth of fifteen feet. Assays of this ore returned $80 in gold and silver per ton.

TRIUNE, JESSIE AND OCCIDENT.

These three claims promise to equal, if not excel, as producers any three claims in the district. The mineral is in each, and undoubtedly in very large quantities. All that is needed to bring the three into line as great producers is capital expended judiciously in their thorough and practical development. Outside advantages are all that could be desired. An excellent mill site is contiguous to all three properties. Timber suitable for mining purposes abounds, and the water supply is abundant. I shall first report upon the leading member of this triumvirate, which has been named

THE TRIUNE MINE.

This claim is located on the east side of Lake Wanicutt, about 300 feet above the mining camp of Golden, and distant from the lake about 1,500 feet. It was located on the 13th day of August, 1887, by Messrs. George E. Darby, of Spokane, and Charles H. Schepster, Charles Cole, and W. H. Townsend. Work of ascertaining the value of the property was commenced forthwith.

The vein at the croppings, at one point, was found to measure 120 feet in width, and the average width of croppings to be fifty feet between walls. The ore is white quartz, carrying about two per cent. of sulphuretts. The surface workings consist of a shaft down thirty-six feet. At the bottom of this shaft a cross-cut has been run a distance of fifty feet from wall to wall. Another shaft is situated near the center of the claim. The depth of this shaft is nineteen feet, all in ore. The quantity of exposed ore on the surface is simply enormous.

Average assays of Triune ore, $6 per ton of free gold. A sample lot of 150 pounds of the ore, worked in a hand mortar, gave a return of $26.50.

The Jessie has a vein three and one-half feet thick of gold milling quartz, and is situated 400 feet further up the mountain from the Triune. The croppings are exposed a distance of 800 feet along the property. The ore is the same as that of the Triune. The Jessie belongs to Messrs. Schepster, Townsend and Darby, who also own the Occident claim, a location adjoining the former. The vein on the Occident is six feet in width—quartz, carrying sulphuretts, gold and silver. On the Occident is a tunnel run in ore for a distance of seventy-five feet. On the two claims (Jessie and Occident) are about 200 tons of ore on the dump. Assays from the Jessie croppings averaged $18 gold, one and one-half ounces silver, per ton. Assays from Occident croppings, $10 gold and two ounces silver, per ton.

THE SPOKANE CLAIM

Is the south extension of the Triune mine. The vein is of quartz and iron pyrites, carrying gold and silver. A tunnel forty feet in length has been run in on the vein, and a shaft sunk a distance of twenty feet in ore. About fifty tons of ore are on the dump. Assays of samples of ore from tunnel, shaft and dump returned:

Minerals.	*Tunnel.*	*Shaft.*	*Dump.*	
Gold, ounces	8	7½	7½	Per ton, 2,000 lbs.
Silver, ounces	2	2	2½	

THE SILENT FRIEND

Is the suggestive title given by its discoverers to a claim located east of the last named claim. Here is a large vein of quartz and iron pyrites. A shaft sunk to a depth of thirty feet on the vein comprises the work completed on the Silent Friend. Assays of ore taken from the croppings returned, silver $10, gold $2, per ton.

Other prospects, on which but a small amount of work has been done, are the Sailor Boy, Great Northwest, Wide West and others. Among the most promising of these are the prospects of the Wehe Brothers, consisting of twenty-seven claims located on Palmer mountain, east of the Ivanhoe mine. The principal claims of Messrs. Wehe are the Pasco, Alyna and Adelia.

The latter claim is situated in a slate formation, and its product is a high grade gold and silver ore. The vein is about five feet in width. Several shipments of Adelia ore have been made to smelters with good results.

George Dance and John Shannon, of Seattle, own the Fairview, Kingbird and Grubstake prospects. The development completed on this group comprises a shaft down thirty feet on the Fairview, a shaft down ten feet on the Kingbird, and one down twenty feet on the Grubstake claim. Three separate veins parallel each other here, each carrying a similar class of ore, it being a quartz and sulphuretts, carrying gold and silver. Assays of the ore from these mines returned:

Minerals.	*Fair-view.*	*King-bird.*	*Grubstake.*	
Gold	$10	$12	$10	Per ton, 2,000 lbs.
Silver, ounces	2	5	3	

George W. Yancy, of Seattle, owns the Independence claim, and Frank Waterman the Leadville, both of which are worthy further development, as are also the Helena and Goldfinch claims, which two latter are now being developed under direction of Col. W. B. Moore.

South of Palmer mountain, in this district, is Chilson's hill, on which are located several fine-looking properties. These are owned by George W. Yancy, of Seattle, and Messrs. Chilson Brothers, of Conconully.

THE ALTA CLAIM,

One of these properties, has a vein of six feet of galena ore, carrying gold, silver and lead. A shaft has been sunk on the vein to a depth of twenty-five feet, and there are about ten tons of ore on the dump. Assays of samples from shaft and dump returned:

Minerals.	*Shaft.*	*Dump.*	
Gold	tr.	tr.	Per ton, 2,000 lbs.
Silver, ounces	90	87	
Lead, per cent	60	42	

The Santa Anna is an extension of the Alta claim, and upon it development will comprise about fifteen feet. Chilson Brothers and Yancy also own the Jumbo and Samson claims. On these only the requisite amount of assessment work has been completed.

MT. CHAPAKA DISTRICT.

It was in the Mt. Chapaka mining district of to-day that some of the earliest gold discoveries were made in the state. When this district was part of the government reserve alloted Chief Moses and his tribe, and when military forces were there stationed to guard against invasion of the domain by the white people, venturesome prospectors eluded their vigilant watch, crept stealthily up around Mt. Chapaka, prospected thereabouts, found the country rich in mineral resources, and, returning to civilization, so reported. The natural sequel was the result. There was a rush to "Mt. Chapaka," as the territory, covering an area of nearly 5,000 acres, was generally designated. A camp was soon in order, and, full-fledged, governed in regulation mining camp style; this gave promise of being an important center in that far-off region, when one morning the military put in an appearance. The result of the visit of the regulars may be imagined. What bid fair to be a town was deserted in twenty-four hours. Inside of a week there was not a prospector on the reservation. It was not until 1875 that the conditions admitted a return to Mt. Chapaka and its environs by the miners. As a result of research begun and continued from where their predecessors had been forced to relinquish it, Mt. Chapaka mining district was formed July 4, 1876, the Hon. H. F. (Okanogan) Smith being one of its founders. There are half a dozen good properties in the district as a result of a small but thorough measure of development given its mineral resources. Ranking prominent among these is the

JULIA CLAIM.

This is a silver and gold producer, situated on the east side of the Similkameen river. The vein here is seven feet in thickness. The hanging wall is of mica schist, there being no foot wall. The ore is a quartz, carrying galena, and of it about 35 tons are on the dump. Assays of samples taken from the dump returned from $100 to $150 per ton. A shaft has been sunk on the property to a depth of 150 feet, and a tunnel driven a distance of 118 feet. The Julia is the property of John McDonald and Hon. H. F. (Okanogan) Smith.

MESSRS. McDONALD AND SMITH

Are also owners of eight mining claims, properly designated the "Eureka group of mines." The group comprises the Ellemehan, San Francisco, California, Kelley, Pontiac, Utica, California Cross Course and Big Tyee claims. The first named is a meritorious prospect. The ledge varies in width from three to seven feet, and on the surface may be traced nearly 1,500 feet. A shaft has been sunk on this claim to a depth of seventy-five feet. A cut has also been completed, exposing to view a deposit of ore estimated to contain about 500 tons.

On the Ellemehan a considerable measure of development has been completed, in the shape of tunnels, cross-cuts and drifts. The Ellemehan has a well-defined vein about five feet in width. Other members of the group show a greater or lesser degree of development. The Eureka group is in process of patent, and this granted, its members, I presume, will be thoroughly developed.

Assay returns from samples of ore from the members of this group are given below.

Mine.	*Gold, ounces.*	*Silver, ounces.*	*Lead, per cent.*
Ellemehan	tr.	15	6 (Per ton, 2,000 lbs.)
San Francisco	tr.	15½	6½
California	.10	14	7
Kelley	tr.	18	6¾
Pontiac	tr.	18½	8
Utica	tr.	15	7¼
California Cross Course	tr.	15¾	9
Big Tyee	tr.	30	6¾

THE CHAPAKA CLAIM

Is a claim named after the district in which it is located, and is situated on the eastern slope of Mt. Chapaka, and formerly belonged to Hon. Henry Hamilton, of Conconully. The vein is ten feet in width, carrying gold and silver ore. A tunnel has been driven on the vein about thirty feet, and further development is being pushed with the aid of three shifts of men, under the direction of Prof. Riley, a well-known mining man. The vein matter is a quartz, carrying iron and copper sulphides. Considerable ore is on the dump. Assays from samples from tunnel and dump returned:

Minerals.	*Tunnel.*	*Dump.*	
Gold, ounces	1.10	.80	Per ton, 2,000 lbs.
Silver, ounces	75.00	125.00	
Copper, per cent	1.50	1.90	

THE CAABA CLAIM

I described at length in my first annual report. I deem here a description of its location and additional development given it all that is necessary. On the Caaba an inclined shaft has been completed a depth of 100 feet through ore. The Caaba still belongs to the Blinn estate, of Portland, and Hon. H. F. Smith. There are about 500 tons of ore on the dump. The Caaba is essentially a concentrating proposition. Assays of ore taken from croppings, shaft and dump returned:

Minerals.	*Croppings.*	*Shaft.*	*Dump.*	
Silver, ounces	22	18	20	Per ton, 2,000 lbs.
Lead, per cent	8	8	8	

THE EAGLE MINING COMPANY

Owns a claim on the east side of the Similkameen river which they have called the Eagle mine. This claim has a vein of quartz, carrying silver and lead, twelve feet in width at the cropping. A tunnel has been driven a distance of 110 feet, tapping the ledge at a depth of 300 feet. A cross-cut between the walls shows the vein to be twelve feet in width at this depth.

Assays of samples of ore taken from croppings, tunnel and breast returned:

Minerals.	*Croppings.*	*Tunnel.*	*Breast.*	
Silver, ounces	18	32	310	Per ton, 2,000 lbs.
Lead, per cent	10	12	....	

THE WASHINGTON

Mining Company is the possessor of a quartette of fine looking claims which they have designated the Diana, Enterprise, Olympia and Highland. This corporation hails from Olympia. All these claims, as far as developed, show good veins of silver and lead ore.

THE DELAWARE CLAIM.

Is located on the south side of the Similkameen river. It is the property of Messrs. Hamilton, Smith and Gibson, and is a silver and gold proposition. The vein is a quartz, carrying silver glance, and is two feet in width. Two shafts, each sunk to a depth of fifteen feet in ore, comprise the work completed on the Delaware. Assays of ore from croppings, each shaft and dump, returned:

Minerals.	Croppings.	Shaft No. 1.	Shaft No. 2.	Dump.
Gold	.10	.20	.10	1.10 Per ton, 2,000 lbs.
Silver	110.10	115.00	111.00	114.00

Among many prospects undergoing at the date of this report a greater or less degree of development are the Dexter, State of Maine, Leviathan, Hunter No. 1, Comstock, Oswego, Fairview, Napoleon, Jenny Lind and Okanogan. Of these the Dexter shows the most development. This claim is owned by M. S. Dudley, Thos. Hanway and Frederick Bowman. It is located on the west side of Mt. Ellemehan, one and one-half miles from Palmer lake. The ore is a quartz and silver glance. A drift on the ledge has been completed a distance of 300 feet. Assays of the ore have returned as high as $300 in gold and silver.

GALENA MINING DISTRICT.

This district was organized June, 1886, by Messrs. H. A. DeHaas, L. Benoist, the Chilson brothers and Col. Wm. B. Moore, and lies to the eastward of the Salmon river district and the Okanogan river. The lack of attention paid this district in 1890, and that attracted toward it the past year, correspond. Why this is the case it would be difficult to determine. Claims reported upon by me a year ago remain under the same control, but the measure of development accorded these, like other properties, has been uniformly small. Good looking claims are the Sonoma, Silver Belle, Nevada, Gussie Everett, and Sunnyside. While the Kismet and Silver Side also look promising.

THE SILVER BELLE CLAIM

Was located by H. A. DeHaas March 26, 1888. This claim is in the northern part of the district. The vein is of quartz, carrying silver and copper glance, two and one-half feet in width. Expense of development thus far completed will approximate $400. Ben. R. Everett and W. H. Cogswell, of Tacoma, now own the claim.

The Silver Side is an extension of the Silver Belle, and is owned also by Messrs. Everett and Cogswell. Here about $300 have been expended in development work.

The Gussie Everett and Kismet complete located extensions on the Silver Belle lead. Both are owned by the same parties.

Northeast of these two claims are two prospects called the Ben Everett and Yellow Jacket. The Lulu, Eureka and Black Huzzar claims are northwesterly extentions of the Kismet vein. Gus. Lieder and A. D. McKey own these claims. On the Lulu is a vein four feet in width of quartz, carrying silver and copper glance. On the Eureka the vein is also about four feet, with the same character of ore. The Black Huzzar carries the same class of ore in a vein three and one-half feet in width. Assays of samples of ore from each of these claims returned:

Minerals.	*Lulu.*	*Eureka.*	*Black Huzzar.*	
Gold, ounces	.10	.10	tr.	Per ton, 2,000 lbs.
Silver, ounces	320.00	370.00	90.00	
Copper, per cent	.40	3.90	3.50	

Isaac H. Durboraw, of Tacoma, owns the Gold Cup, Gold Eagle and Virginia claims, contiguous to the above described properties. The sum of $500 will cover expenditures thus far made for development work on these properties. Assays of ore from the claims returned as follows:

Minerals.	*Virginia.*	*Gold Cup.*	*Gold Eagle.*	
Gold	$30	tr.	tr.	Per ton, 2,000 lbs.
Silver	5	$80	$120	

SILVER BLUFF CLAIM.

The Silver Bluff location is that of Chilson Brothers, of Conconully, and H. McCartney, of Salt Lake city. The vein is two feet wide on the surface. An inclined shaft has been driven sixty-five feet in ore. The ore is a quartz, carrying copper and silver glance. About fifty tons of high grade ore are on the dump. Assays of samples from croppings, inclined shaft and dump returned:

Minerals.	*Croppings.*	*Shaft.*	*Dump.*	
Gold, ounces	tr.	tr.	tr.	Per ton, 2,000 lbs.
Silver, ounces	110	97.00	102.00	
Copper, per cent	3	3.50	2.90	

The King Solomon, a prospect owned by the same parties, is southwest of the above; and among other prospects are the Occident, Palmetto, Pole Pick, Sonoma and Union.

THE CHLORIDE DISTRICT.

This district is but two years old, having been organized in 1889, by ex-sheriff Robert Allison, of Okanogan county, Hill Thomas, Thomas Dixon and others. The chloride ores predominating in the district caused the title to be given it. This district lies southwest of Ruby district, and in the western portion of Okanogan county, between the Salmon river and Methow district Argentite (vitreous silver, or silver glance), stromeyerite (a silver-copper glance), acouthite (silver sulphide), chalcocite (copper glance, carrying silver), proustite (light red silver ore), tetrahedrite (gray copper ore), and pyrargrite (ruby silver ore), rank among the various mineral ores in this district. It is new, and but little in the way of extensive development has been yet accorded any of the claims located within its boundaries. Principal among the claims is

"THE CHLORIDE,"

A location of ex-sheriff Robert Allison, H. E. Davis and John Mulholland, of Okanogan county. This claim has been accorded a fair measure of development work. A tunnel 100 feet in length has been driven in on a vein of white quartz, carrying silver glance, brittle silver and copper carbonates. This vein is four feet in width and has been uncovered on the surface in several places. Assays of ore from surface croppings, tunnel and dump returned:

Minerals.	Croppings.	Tunnel.	Dump.	
Gold, ounces	tr.	tr.	tr.	Per ton, 2,000 lbs.
Silver, ounces	245.00	650.00	315.00	
Copper, per cent	1.25	2.00	1.50	

HILL, THOMAS AND OTHERS

Own a claim, the vein of which runs northeast and southwest, dipping 20 degrees to the west, in which there are thirteen inches of ore. The work completed comprises an inclined shaft in a distance of seventy-five feet. On the dump are about thirty tons of ore. The ore, in character, is similar to that found in the Chloride mine. Assays of samples from croppings, shaft and dump returned:

Minerals.	Croppings.	Shaft.	Dump.	
Gold, ounces	tr.	tr.	tr.	Per ton, 2,000 lbs.
Silver, ounces	118	320	325	

Chloride prospects of promise include the Moore group, Roseman and Haggerty groups, and the La Plata claims.

THE METHOW MINING DISTRICT.

This district is the largest, if least developed, of any of the mineral subdivisions of Okanogan county, and is located in the southeastern corner of the county.

The Methow takes its name from the Methow river, quite a large sized stream, traversing its area, and embraces a vast domain, with the river its principal water course.

Why the Methow country has not yet attracted the attention of capitalists and mining men to a far greater degree than it has, I cannot understand. The country, from what is thus far known of it and its resources, from a mineral standpoint, certainly is deserving of a greater share of attention, and incident thereto, more careful and systematic investigation than has thus far been accorded it.

Aside, of course, from railroad transportation, a most essential, yet sadly lacking requisite in all our state's mineral regions, the Methow may be classed among the most of the many naturally accessible mineral subdivisions of the state. The country is easily approached, and is, from a topographical standpoint, one of the most open in the Northwest.

With the fact demonstrated beyond peradventure that the Methow's mineral resource is as permanent and valuable as present appearances certainly indicate it to be, it will be found a veritable paradise for both prospector and development worker. On its surface will be found a practically unlimited supply of timber, capable of being manufactured into everything needful of the character for mining purposes and for architectural uses. The species are as varied as the quality. Fir, spruce, tamarack, mountain pine and larch maple are there to be found, and the forests of the Methow will be found remarkably free from early or more recent visitations by fire.

Water in abundance may be found throughout the country, and any number of excellent mill sites are contiguous to prospects already located in the mineral belt.

The Methow's climate is excellent, being remarkably equable and healthful. While in its latitude naturally subject to the rigors of a winter season, the chilling blasts experienced by residents in the plains country, even far to the southward of the Methow, are there unknown. Surrounded by mountains, the country is sheltered to a degree, and habitation the year round in the Methow will be found congenial and comfortable. Contiguous to the Methow is a rich agricultural, as well as, supposedly, a mineral district. The Lake Chelan country lies to the south and west of it. The fruit belt of Okanogan is at its very doors, and in the Methow itself are some of the very best of agricultural lands, awaiting only the coming of the farmer and the introduction of the plowshare.

Along the Methow river, Gold creek and the respective tributaries of these streams are located the principal claims of the district, using the term in so far as its application to the amount of completed development work may be applied.

Of these there is the Red Shirt claim, undergoing a fair measure of development at the hands of a Montana syndicate of mining operators. The ore in character is an arseno-pyrite (mispickel ore). On the vein a tunnel has been driven a distance of 125 feet, and a promising deposit unearthed. Some thirty tons of ore are on the dump as the result of work completed ou the claim. Assays from samples of the ore taken from croppings, tunnel and dump returned:

Minerals.	*Croppings.*	*Tunnel.*	*Dump.*	
Gold	$8	$30	$22	Per ton, 2,000 lbs.
Silver, ounces	15	15	16½	

THE CRYSTALITE AND GOLD CREEK

Are claims situated just across Gold creek from the old "Stemwinder" location named in my last report. These claims were located by John Runnels and Thomas Deaver, of Conconully. These gentlemen began developing the former claim in the early part of last summer, and as a result have demonstrated the Crystalite worthy of thorough examination. The ledge is two feet in width, carrying a high grade of antimonial silver, black sulphuretts and oxide of iron and copper ore.

A short surface cross-cut has been completed. The claim might be better described as located about fifteen miles above the mouth of, and five miles from the bank of, the Methow river. On the Gold Creek but the requisite amount of assessment work has been completed. The Crystalite ore runs from 50 to 350 ounces to the ton in gold and silver.

These parties also own the Sidewinder claim, opposite and across the river from the Crystalite and Gold Creek prospects. The Sidewinder ledge is four and one-half feet in width, and the development work completed this year approximates twenty-five feet. The ore is similar in character to that discovered in the Crystalite and Gold Creek, but not so rich in mineral.

LAKE CHELAN DISTRICT,

So named from one of the most beautiful of lakes in Washington, and located in the southeastern portion of Okanogan county, is one of the many new districts formed the past year.

While the advantages of soil and climate of the "Lake Chelan country," as it is generally called, have been known for several years as a possible seat of mineral wealth, that region has only just begun to attract attention. That it lies contiguous to one of the richest mineral divisions of Washington, is the fact. That the limited research given it by prospectors has thus far been attended with most encouraging results, is also the fact. I look for good reports from this new district the coming year.

Like the Methow, the Lake Chelan country offers, as far as surface advantages are concerned, every inducement to the prospector. There is abundance of timber, water and game, and the products of the soil embrace everything of value to the agriculturist. The ores found are gold, silver, lead and copper.

Chelan district was formally organized by miners and citizens of Chelan and vicinity August 22, 1891, and the boundaries set forth as follows: Commencing on the Columbia river at a point of rock two miles east of the Antwine place; thence running in a northwesterly direction, following the divide between the Methow river and Lake Chelan, to a junction with the Skagit county line; thence in a southeasterly direction, following the divide between the Entiat river and Lake Chelan, to the Columbia river; thence along the west bank of the Columbia river to the place of beginning. John Carlisle was elected recorder of the district, and a set of by-laws governing it, which are in consonance with the mining laws of the United States, was adopted.

STEVENS COUNTY.

ANOTHER LARGE MINERAL DIVISION OF THE STATE.

The county of Stevens, since the issuance of my first annual report (Mines and Minerals of Washington, 1890), has forged rapidly toward the front as one of the most important counties in the state, both as regards its practically unlimited resources of a mineral character, the agricultural value of its prolific soil, and the attention attracted to its splendid forests by mill and lumber men. Within the past twelve months a truly wonderful improvement may be said to have taken place within this county from an economic standpoint. Despite a general commercial depression prevalent in the state the past year, and which is now happily giving place to an era of unbounded prosperity and practical development, Stevens county seems to have kept fully apace with that measure of advancement characteristic of the state in general in the previous year.

Stevens county, as regards that prime essential to both rapid and thorough development in any region — accessibility by modern methods of transportation, may be said to have been more favored at the outset than most of what must, at present, be designated our counties remote from the leading commercial centers of both this state and adjacent territory. Stevens enjoys direct rail communication with all of Washington's marts of trade and manufacture. The county may aptly be described as bisected by a thoroughly constructed standard gauge line of railway; and enjoys the possession of a stream of water already navigable for a considerable distance, and which may be made so almost to the northern boundary of the county at comparatively small expense.

Geographically, Stevens county lies in the northeastern corner of Washington, it in fact occupying the northeasternmost portion of the state, being bounded on the north by the British Possessions (the famous Kootenai mineral region); on the east by the northwestern portion of the State of Idaho; on the south by the county of Spokane, and on the west by the reserve set apart by the United States government for the use of the Colville Indians. The most popular means of ingress and egress is via the line of the Spokane Northern Railway, which, as I have stated, practically bisects the county. Its area may be easily reached from eastward of its boundaries by good wagon roads, which also enter the county from the British Possessions, the south and the west.

TOPOGRAPHICAL.

The topographical aspect of the county, from a position in its southeastern portion, is a succession of undulating country, gradually assuming a higher altitudinal plane toward the northwest, and presenting an appearance, on the whole, of a surface gradually evidencing more pronounced upheaval northward, and terminating in a succession of lofty peaks and spurs of ranges extending on into the British Possessions. The undulating, and later, more pronounced upheavals, are cut at about the median line of the county (north and south) with a large valley called the Colville valley, and widely known for the fertility of its soil. At intervals, in the undulating and more upland area, are found small and fertile valleys, and in the mountainous area stretches of fine grazing and agricultural lands of greater or less extent.

THE COUNTRY ROCK OF STEVENS

Is granite, syenite, porphyry, slate and lime. The different characters of ore found comprise horn silver, ruby silver, carbonate of lead, red oxide of copper, purple copper, green copper, blue copper, brittle silver, carbonate oxide and sulphide of iron.

There are many mines of prominence in the county, among them the widely known

OLD DOMINION MINE.

This property was discovered in the month of April, 1885. The Old Dominion, since its discovery, must be credited with an output of ore valued at over $500,000.

The ore of the value stated was stoped from the surface workings. There is still an immense amount of ore in sight. Over 4,000 tons have been shipped, averaging 200 ounces in silver and thirty per cent. of lead, since the opening of the mine. The pay roll has been as high as $5,000 per month, and at the present time over $1,000 per month is disbursed.

The property is located on what is known as Dominion mountain, Colville mining district, six miles east from the town of Colville, on the Spokane & Northern Railroad, near the center of Stevens county. The ore is a quartz, carrying brittle and ruby silver, copper, carbonate, and sulphide of lead, and gold. The ore is in quartzite and lime. The management is engaged in running a tunnel 1,000 feet. The tunnel is at present in 925 feet, the object of driving being to strike the main ore body at the depth of 500 feet below the old workings. On the dump of the mine are about 3,000 tons of ore that will average about thirty ounces in silver per ton, beside what is in the slopes of the old workings. I have succeeded in procuring some exceedingly rich specimens of ore from this mine. The mine can be easily reached by a good wagon road from Colville and the railroad. Its surface equipment is of the best. I deem the Old Dominion about as thoroughly and systematically operated a mine as could be found in the Northwest. A concentrator will be in operation in March at the mine, which is under the general management of Col. Field, an experienced miner from Colorado.

THE DAISY MINE.

I deem next in importance as a mineral proposition a property situated in Summit mining district, in the western part of the county, within about two miles of the Daisy post office, on the Columbia river, sixteen miles south from the Spokane & Northern Railroad. I base my reasons for so doing on the mineral showing at hand and the fact that it is made easily accessible from the railroad and post office over a good wagon road, and the important feature that from mine to rail communication is a down hill pull over an equally excellent thoroughfare.

The Daisy mine is a carbonate ore proposition, located at an altitude of 1,500 feet above the Columbia river, in the heart of one of the richest mineral producing districts of Stevens county. It has been only within the last two years that active development work was begun on the discovery. During that period sufficient demonstration has been given its owners that in their mine they have struck a veritable Drum Lummond, and I must confess my concurrence in their belief, after a close inspection given the property.

The ore in this mine at the surface has been cross-cut, showing the vein to be over forty feet in width; and a drift 200 feet long has been run through ore. A shaft has been sunk sixty-three feet, showing ore all the way. The average of this ore, as shown by assays, returned thirty ounces silver, twenty-five per cent. of iron, and eighteen per cent. of lead. At a distance of sixty-three feet below the upper workings a tunnel has been driven a distance of 165 feet to connect with the shaft to which I have alluded. From this shaft a drift has been run in on the vein a distance of 310 feet. The vein at this point averages nine feet in width, and assays about the same as the surface ore. Fifty-five feet below the tunnel referred to, another level has been driven eighty-six feet, tapping the vein and running along the vein 250 feet to connect with the shaft. At a distance of 145 feet below the fifty-five foot level a tunnel has been driven 409 feet tapping the vein, and a drift has been driven 165 feet in ore, and wants but a few feet to connect with the shaft from the apex of the vein. As will be noted, I have blocked out the ore body, and estimate that 30,000 tons of ore are in sight.

Over 4,000 tons of a desirable smelting ore have been shipped from the mine to the various smelters in the west, the Tacoma Smelting and Refining Company receiving the lion's share.

THE SILVER CROWN MINE

Is a silver and lead proposition located in the Little Dalles district, in the northern portion of Stevens county, on the Columbia river, about five miles from Little Dalles post office and within 500 feet of the Spokane & Northern Railroad. It is the property of the Silver Crown Mining Company, which also owns the Northern Light mine, described below. The Silver Crown is being worked by a large force of miners under the direction of Mr. A. K. Kelley, an experienced miner and the general manager of the company.

I found at this mine considerable development work completed and a large amount in hand. A sixty-foot shaft has been sunk. A tunnel has been driven 130 feet, and cross-cuts completed aggregating seventy-three feet. A tunnel driven 100 feet, from a point 200 feet below the surface, taps the vein.

The vein is three feet wide, of solid galena carbonate and oxide of lead, and lies in quartzite and lime. The strike of the vein is east and west, being almost vertical. On the dump I measured about 150 tons of ore. Average assays of this ore returned ninety ounces of silver and forty per cent. of lead.

On the Northern Light a tunnel has been driven a distance of seventy feet and a shaft sunk forty feet through ore of similar character to that in the Silver Crown. About 3,000 tons of ore are in sight. A tunnel, to be driven 600 feet, has been commenced, to tap the veins of the Silver Crown and Northern Light at a depth of 750 feet. On completion of the tunnel the company should be able to ship from forty to fifty tons of ore per diem, and considerably increase its working force.

THE YOUNG AMERICA MINE.

This mine was discovered in October, 1885, and is operated by the Young America Consolidated Mining Company. It is a smelting proposition, and the ore is galena. The vein is five feet wide, bearing northeast and southwest, and lying in lime. About 250 feet of development work in the way of shafts and tunnels has been completed. A shipment of a few car lots of the ore has been made to different smelters. One shipment returned ninety-three ounces of silver and thirty-seven per cent. of lead. About 150 tons of ore were on the dump when I last visited the mine, and assays from this ore returned eighty-two ounces of silver and forty-seven per cent. of lead per ton.

The Young America is near the Columbia river, in Colville mining district, about sixteen miles north of the town of Colville, on the Spokane & Northern Railway. The mine may be reached from both town and railway by a good wagon road.

THE BONANZA MINE.

Belonging to the Consolidated Bonanza Mining and Smelting Company, is located near the Young America mine and distant from the town of Marcus, on the Spokane & Northern Railway, about seven miles. The mine was discovered in October, 1885, and until the incorporated company named secured the property it was operated by Messrs. Marcus Oppenheimer, C. H. Armstrong and H. Ensler.

After an expenditure of $1,500, the owners succeeded in extracting about 2,000 tons of ore, which was shipped to a smelter and handsome returns realized therefrom. Average assays of the ore were from fifty to sixty-five per cent. lead per ton. My assays of ore from this mine showed eighteen ounces of silver and 62.50 per cent. lead per ton. Development completed comprises two cross-cuts, one of sixty and the other forty-two feet in length. A shaft is down thirty-five feet. I found the mine had several thousand tons of ore in sight. The surface improvements comprise whim house, boarding house and shaft house. A good wagon road connects the mine with the main county road.

The new ownership represents a capital of $1,000,000, and it is proposed to erect a smelter near the property the coming spring, at which both Young America and Bonanza ores will be treated.

THE EAGLE MINE.

This is the property of the Eagle Mining Company and is situated in the Chewelah mining district, in the center of Stevens county, and sixty miles north of the city of Spokane. The mine is within one and one-half miles of the town of Chewelah, on the Spokane & Northern Railroad. The output of this mine is a smelting ore of cerussite or carbonate of lead, and anglesite or sulphate of lead, oxide of lead and galena. The vein is about four and a half feet in width, with a trend toward the northeast and southwest, dipping to the southwest. It is a contact vein in quartzite and limestone. This mine was discovered in 1885 by Wagner, Henshaw and Williams, and was sold by them to the present operators, the Eagle Mining Company. About 1,600 feet of development work has been completed in the shape of tunnels, winzes, shafts, etc. An excellent steam hoisting plant is on the property, and a shaft is being sunk 250 feet deeper than the present workings. On an average, twenty-five men are employed in and about the mine. The pay roll amounts to about $2,500 per month. The company has been shipping about fifteen carloads of ore a month to smelters in Butte and Tacoma, the output averaging about ten tons per day. On the dump I measured about 200 tons of ore. Assays of the ore from this property made by me returned about forty ounces of silver to the ton, with fifty-two per cent. lead.

THE EXCELSIOR CLAIM.

This claim was located in the spring of 1886, and is owned by J. H. Young & Co. It is located near the Silver Crown Mining Company's property, in the Little Dalles district, near the Spokane & Northern Railroad. This is a galena ore producer. The vein, two feet in width, carries iron pyrites, and runs east and west, dipping into the hill. A tunnel has been driven into the hill, tapping the vein. About thirty-five tons of ore were on the dump on the occasion of my last visit to the property. Assays of the ore on the dump made by me gave the handsome returns of 118 ounces of silver, fifty per cent. lead and fifteen per cent. of iron.

THE SUMMIT CLAIM

Is located in Chewelah mining district. It is situated about eighteen miles west of the town of Chewelah, a flourishing camp, and is directly on the line of the Spokane & Northern Railroad. A good wagon road renders the mine easily accessible from both Chewelah and the railroad. J. N. Squier owned the Summit mine, but disposed of the same during the past year. It is one of a quartette of fine properties formerly owned by him and designated the Squier group of mines. These I shall describe *seriatim.*

The Summit is a galena and anglesite (sulphate of lead) proposition. The vein runs north and south, dipping to the westward. It is three feet in width. The Summit has a shaft sunk to a depth of 300 feet, with a level extending fifty feet from the bottom of this shaft, tapping the vein. The shaft is nicely timbered, and one of the most thoroughly constructed that I have seen in that section of the mineral belt. On the dump I measured seventy-five tons of first-class ore of the variety I mentioned above; and also 200 tons of second-class ore. The latter returned fifteen onnces of silver and twenty per cent. lead per ton. Assays made by me in the state laboratory from samples taken from the Summit dump gave 63 ounces of silver and 41.30 per cent. lead per ton. This was the first-class ore.

The second member of the Squier group is known popularly as the Blanche lode, which is situated close to the Summit mine. The ore is a quartz and copper glance, of which there is a two-foot vein. Assays taken by me have returned 100 to 400 ounces of silver, and eight to ten per cent. copper, per ton.

About thirty tons of ore were on the dump on the occasion of my last visit to the property. Several carloads of high grade ore have been shipped from the mine to the Denver smelters. A fifty-foot incline has been driven through ore. The surface improvements are of a substantial character.

On the remaining two claims in the Squier group comparatively little development work has been done.

BELLE OF THE MOUNTAIN CLAIM.

In Summit district is situated the Belle of the Mountain mine, a gold producer. Here a vein of gold quartz ore three feet in width has been developed. The trend of this vein is east and west, dipping into the mountain on which the claim is located. About ten tons of ore were on the dump. A shaft twenty feet in depth has been sunk. Assays from the ore on the dump returned me $20 in free gold. This mine is the property of Mr. C. H. Bodge, who has been directing the work of developing the property.

The Blue Bell, of which the Belle of the Mountain is an extension, is situated directly south of the latter property. The strike and dip of the two veins are identical.

The Blue Belle is a gold producer, having a well defined vein of gold quartz ore three feet in width. On this property a shaft forty feet deep has been sunk through ore. About fifty tons of ore were on the dump, from which I took samples promiscuously, which returned me $20 per ton in free gold. P. Kearney, D. B. Arman and J. Davis are directing its development. The prospect of becoming a rich gold producer I think flattering for this mine. Both the Belle of the Mountain and the Blue Belle are rendered easily accessible from camp and railway by good wagon roads, and could be made to produce fifteen tons of ore per diem, enough to warrant the erection of a ten-stamp mill.

THE WELLINGTON CLAIM,

Located in Summit district, four miles south of the Daisy mine, heretofore described by me, is the property of W. H. Kearney, the manager of the Old Dominion mine, of which I have also written. The Wellington is essentially a smelting proposition, the ore being a carbonate of lead and iron. A shaft has been sunk a depth of thirty feet on the vein, which is two and a half feet in width. A cross-cut has also been run; both shaft and cross-cut in ore. On the dump I measured about twenty-five tons of ore. From assays taken I received returns as follows: Seventy-five ounces of silver, twenty per cent. lead, and thirteen per cent. iron, per ton.

THE VICTORY LOCATION

Is a very promising claim, located in the Summit district, being a parallel claim to the Daisy, lower down the mountain. It has a four-foot vein of quartz, carrying galena, iron pyrites, etc. The development of this property has, thus far, not been of an extended character. A shaft fifteen feet deep, in ore, comprises the work completed. On the dump I measured about twenty tons of ore. From assays taken from the dump I got as returns sixty ounces of silver, thirteen per cent. lead, nine per cent. iron, per ton (2,000 pounds). If this mine was properly developed, I am of opinion that its output per day could easily be made ten tons.

THE OLD ABE LOCATION,

The property of Mr. J. N. Squier, a pioneer prospector of Stevens county, and owner of the Squier group of mines I have described, is situated in the Summit district near the Columbia river, and transportation by rail and water. This is a smelting proposition. Here a four-foot vein of quartz carrying iron and copper has been developed. A forty-foot shaft has been sunk on the vein, traversing that distance through ore. At the bottom of this shaft, for a distance approximating fifteen feet, a drift has been driven. On the dump I measured twenty tons of ore. I secured valuable samples of this ore, and at the state laboratory I assayed them. My returns were: Twenty ounces of silver, five per cent. copper, fifteen per cent. iron. With proper development this property, like the one above described, should be made to produce ten tons of ore per diem.

THE CAPITAL CLAIM.

This property is situated about eight and one-half miles south of the town of Chewelah, on the county road, in Chewelah mining district, within easy access of both town and railroad over a good wagon road. The mine is the property of E. E. Alexander and H. P. Reeves, of Stevens county. The vein here is seven feet in thickness, of limonite iron or brown hematite of iron. The strike of the vein is northeast and southwest, dipping to the westward. The hanging wall is porphyry and foot wall of lime. Development work comprises a shaft now down twenty feet, through ore, and several small cross-cuts. On the dump are thirty tons of ore. Samples of this oxide of iron taken from the dump gave fifty-five per cent. metallic iron, six and one-half per cent. silica, a trace of manganese, four per cent. of lime, a trace of sulphur, and no phosphorus. This is a most desirable ore, and will be sure to be in great demand.

THE FINLEY MINE,

A fine property, is situated eight miles northwest of the town of Chewelah, on the Spokane Falls & Northern Railroad. It is the property of W. H. Fife, of Tacoma, J. M. Buckley, of Spokane Falls, and W. E. Sullivan, of Chewelah. Here is a three-foot vein of quartz, carrying galena, copper and iron pyrites. The strike of the vein is northeast and southwest, dipping to the north. Developments consist of one inclined shaft sunk through ore on the vein a distance of seventy-five feet. A lot of ore shipped to a Montana smelter yielded ninety ounces of silver and twenty per cent. of lead per ton.

For smelting purposes, when the smelter at Spokane commences operations, if this event comes to pass, its proximity to railroad communication, and the very excellent quality of its output, impels me to believe that the Finley claim will soon rank among the most prominent in Stevens county.

THE TENDERFOOT CLAIM.

Fifteen miles east of the flourishing town of Colville, on the Spokane Falls & Northern Railroad, in the Clugston Creek district, on the south fork of Clugston creek, is situated the Tenderfoot claim. T. D. Hayden, sheriff of Stevens county, and P. D. Grace and M. D. Mahoney, of Kootenai, B. C., are the owners. This is a galena proposition, in lime and slate contact. The strike of the vein is northeast and southwest, dipping east, and is four feet in width. Developments made comprise three tunnels, one fifty feet, one ninety, and one 120 feet in length. At the breast of the 120-foot tunnel they have drifted on the ore vein sixty feet north and forty feet south. A wooden ore chute, fifty feet long, has been constructed for the conveyance of ore from the dump to the wagon road. On the dump I measured about 100 tons of ore. From the samples picked up promiscuously and assayed by me I got returns of fifteen ounces of silver and sixty per cent. of lead. The amount of ore in sight on the occasion of my last visit to the mine would approximate 1,000 tons, and the mine is capable of an output of fifteen tons per diem. Several carloads shipped to the smelters averaged fifteen ounces of silver and sixty per cent. lead per ton. On the surface considerable work in the way of substantial improvements is to be noted.

SILVER LEAD LOCATION.

This property, under the same ownership as the Tenderfoot claim I have just described, is located directly across a gulch intervening between it and the Tenderfoot. It is a large deposit of hematite, and is practically undeveloped. A great amount of ore is in sight on the surface. About 1,000 tons of ore from this deposit have been shipped to the smelter at Colville. The lead, gold and silver in the ore has completely reimbursed the owners for the mining, transportation and the smelting of the same. The ore assayed sixty per cent. metallic iron, and is worth $10 to $15 per ton for smelting purposes, being used as a flux.

Among the other well known galena silver properties in the vicinity undergoing development are the "Dandy" and "Moonshine," and one or two others of lesser importance.

BRUCE CREEK DISTRICT.

Although termed a "mining district," Bruce Creek has never been formally organized. It is situated northeast of the town of Colville, in the vicinity of the Clugston Creek mining district, reference to which I have already made. Among properties I inspected in this Bruce Creek district were those of the Al-Ki Mining Company, which owns the principal properties in the district. This is a Stevens county organization, formed in November, 1890, of which S. Douglas is president, and John Kehoe manager. This company owns five claims, viz.: The Silver Wave, Myrtle, Morning, Ranger and Fraction.

The Silver Wave has a four-foot vein of galena. The Myrtle has a three-foot vein of the same character of ore. Considerable development work has been done on all of these properties, and considerable ore lies on the dump of each of them. Assays from the Silver Wave returned fifty-six ounces of silver and sixty-four per cent. lead per ton. Assays from the Myrtle returned seventeen ounces of silver, $2.50 of gold and fifty per cent. of lead per ton.

Another good property on Bruce creek is the Dead Medicine location. This property is owned by Judge George Turner, of Spokane, and George W. Forster, also of that city. The strike of this vein is northwest and southeast, and dips to the north. The ore is a decomposed quartz, carrying galena. Development work consists of a shaft sunk to a depth of fifty feet, through ore. At the bottom of this shaft the ledge matter is six feet wide, with a pay streak of two feet of solid galena. On the dump I measured about thirty-five tons of ore. Assays taken from the vein showed thirty-three ounces of silver, and sixty-two per cent. lead, per ton. Assays from the dump showed thirty-two ounces of silver, and fifty per cent. lead, per ton.

THE METALINE DISTRICT.

Isolated from surrounding settlements, practically cut off from communication with the outside world, is a mining district, the last I shall describe in Stevens county. This is known as the Metaline district, so called from the immense amount of metal in sight, and is located in the far northeastern corner of that county. This is one of the largest and oldest mineral divisions of Stevens county, and is certainly destined to rank a leading mineral producer of the state, once transportation facilities, both by rail and water, are granted it. The wealth of its mineral possessions taken into consideration, I marvel that either the United States government or our own state government have not paved the way for the gaining of the treasure trove I know to be existent in the Metaline. I prophesy that with the removal of the Little falls, an obstruction in the Pend d'Oreille river thoroughly shutting out communication by that water highway with the Metaline, that district inside of a year would be the goal of hundreds of prospectors, operators, capitalists and investment brokers. These falls, if such they may be properly designated, could be moved for less than $5,000, and an avenue of communication with the Metaline district opened up from Lake Pend d'Oreille and the main line of the Northern Pacific railroad, thus permitting the transportation of the minerals of this district to the various smelters scattered throughout the country.

In the Metaline are immense deposits of lead ore in lime formation, many of them actually measuring fifty feet in width. This ore is of the galena and carbonate variety, and I have made assays of this ore that ran as high as twenty ounces in silver and seventy-four per cent. of lead. One of these assays was from a sample of galena ore from the Bonnie Blue Belle mine, now in process of being patented. Another sample, cerussite, or carbonate of lead, ran forty-one ounces of silver and sixty per cent. of lead per ton. This was only an average sample of the vein, which was twenty-eight feet wide, of a solid lead ore.

The obstructions in the Pend d'Oreille river alone keep Metaline district from affording a field of mineral development and building up industrial enterprises only equaled by those of Colorado, where like deposits exist.

Other claims in Stevens county certainly warranting thorough development are the Capital, Iron King, Amazon, You-Like, and Blue Grouse, and the properties of the Kettle Falls Mining and Development Company.

KITTITAS COUNTY.

CLE-ELUM MINING DISTRICT.

One of the pioneer mining districts of the State of Washington bears the name "Cle-Elum." The title is of Chinook origin, and in English its signification is said to be cold water. The district lies in the northwestern portion of the rich agricultural and mineral county of Kittitas, one of the largest and principal political subdivisions of Eastern Washington. Its westernmost boundary may be said to parallel the rising and undulating ridge forming a minor vertebræ of the great Cascade range of mountains. As well as being a pioneer among its fellows, so to speak, Cle-Elum ranks one of the largest mineral subdivisions the state possesses. Its wealth of mineral resource is as varied as its history is romantic. It has been demonstrated a great treasury of all the metals, precious and otherwise, needful requirements of an advanced and progressive civilization. I do not consider a brief *résumé* of a remarkably interesting history of this district out of place in this report. That comprising its advent into the ranks of mineral producing areas here in the Pacific northwest may be said to date back as early as the year 1837.

It was early in the spring of that year that a company, or rather a small party, of French Canadians, Scotchmen and Englishmen, ostensibly an outfit of the Hudson Bay Company, camped on what was then to the aborigines, and is still to the paleface, known as the Cle-Elum river, a small but beautiful stream taking its source at Fish lake, east of the pass of the Snoqualmie, and, flowing south, losing itself in the Yakima river.

The members of this party, as all such early invaders were, were in search of valuable furs. They had met with unanticipated hardships while wintering near the pass of the Snoqualmie, and quickly as clement weather and the condition of the ground admitted, hastened southward along the Cle-Elum river, bound for the broad Columbia.

At a point on the river distant about twenty miles from the now flourishing town of Cle-Elum, the party camped to recruit. Barter with the red man was soon in order. The captain of the party one day noticed in the possession of an athletic "buck" some curiously shaped weapons, the appearance of the material used in their construction bearing a strong resemblance to the iron fastenings on some of the equipment of his party. His curiosity was aroused. He asked for and obtained permission from the Indian to examine his arms. To his surprise he found the point of the aborigine's fish spear was of iron; that a ring in its stock was of the same metal; that a unique medal worn by the savage was also of iron. The iron, however, as regards its manufacture, was not the handiwork of "the white man." Questioned as to where he obtained material for its manufacture, the Indian said it was taken from the hills near by, "cooked" by them, allowed partly to cool, and then hammered into the shape of spears, medals, hammers and other implements of use to the tribe. Astounded, the captain decided to see this hill of iron. He did so. He found a ledge of iron ore. Time pressed him. He gathered specimens, carried them to the company's fort on the Columbia, and there the first iron ore found in Washington was first exhibited. Samples of both ore and crude materials fashioned from its product were afterward sent to England, and are said to be on exhibition to-day in the British museums. But Washington was an unknown wilderness. Stories of mineral wealth, were they recited, would gain for the entertainer the derision of his auditors. Washington furs were alone the quarry, and the trapper and adventurer would have it that her wilderness was valuable for nothing more. Mr. J. Flett, now of Pierce county, as mentioned in my last report, knew of the existence of iron in Washington as early as 1842.

GOLD, SILVER AND COPPER DISCOVERED.

As equally interesting is the history of the discovery of gold, silver and copper in the Cle-Elum country. To the aborigine also is due the credit of these discoveries which were made in the years 1846,'47,'51, respectively. The sight of a "klootchman" with gold bars pendant from her ears and an anklet of gold above her tawny foot must have indeed surprised old Anton Videau, a hardy Canadian trapper, in the first-named year, when he went to barter trinkets to a camp of Indians on the upper Cle-Elum river. His cupidity aroused, he questioned the Indian woman. As a result he was shown a shower of coarse placier gold that dazzled his eyes. He was a "tillicum" of the Siwashes. They showed him the "yellow ground." Videau ceased to be a trapper. He mined the precious metal, secured several hundred dollars and returned to his home of long ago in Montreal. Videau returned with a party in quest of the golden ground, but it is supposed was massacred, with its members, as war between the Hudson Bay men and the aborigines had been declared. But the story had been told at home, and venturesome townsmen visited the Cle-Elum, returned well repaid for their hardships and dangers endured, and so the story of Cle-Elum as a gold field was made public.

Discoveries of silver and copper in the country were made first in the years I have noted. Each was due to the Indians; and news of each was heralded to their white brethren afar by the hardy pioneers who saw silver and copper in the possession of the red men.

The first practical invasion of the Cle-Elum country in search of precious metals was made in the year 1881 by Messrs. Hawkins and Splaun (Mines and Minerals of Washington, 1890).

These gentlemen were veteran prospectors, having followed that hazardous calling for many years prior thereto. They left what was at that time known as the Peschastin district for a tour toward the northwest, hoping to find diggings more remunerative than those from which they departed. They found these in the Cle-Elum mining district, worked them for a time, and then returned to civilization to tell the tale of discovery. Excitement was of course created, and a rush to the Cle-Elum ensued.

TOPOGRAPHICAL.

Represented topographically, Cle-Elum district may be said to be one mammoth gorge, or great coulee, flanked by high and precipitous mountains on both sides; the source of this gorge or coulee being in the northwestern corner of Kittitas county, close by the pass of the Snoqualmie, and its mouth lying contiguous to Cle-Elum lake, about four miles from the Upper Yakima river, also in Kittitas county, its general direction approximating north and south. No definite or clearly defined statement as to the contour of this gorge is possible, because of the numerous indentations and undulations noted the entire length of its walls.

Traversing the entire distance of this gorge is the Cle-Elum river, a stream of considerable size, which I have, as regards general description, hereinbefore particularized. Undoubtedly this stream and the action of other water in periods remote have accomplished what it is plain to be seen they are still accomplishing—the denudation of the rock of its clothing of earth and *debris*. To this action of the waters in thus denuding the rock is due the exposure of the various mineral ledges there abounding. The country rock of the district is mainly syenitic, granity, hornblendic schists, porphyry and slate, and conglomerates.

CLE-ELUM MINES.

On the east side of the Cle-Elum river, near its source, are situated the principal mines. Prominent among these is the Aurora mine, situated on Lynch's mountain. It was located by John Lynch in July, 1886. The vein is about four feet in width, carrying a free milling gold quartz ore, assaying from $50 to $200 per ton. Development work comprises a tunnel and an inclined shaft. Surface equipment comprises a boarding house, blacksmith shop, and a "Chili mill," or "arastre." Mr. Lynch, astonished at the richness of the ore, proposes applying great zeal to the complete development of his property, and the application of approved machinery thereto.

THE MOUNTAIN SPRITE

Is properly an extension of the Aurora. But very little development work has been done on this property as yet, but what has been accomplished has been resultant in showings, admirable in character.

THE AMERICAN EAGLE CLAIM

Is located west of the Aurora, on Eagle mountain, its altitude exceeding that of either of the properties named. Here a vein six feet in width has been located. It carries a pay streak of about one and one-half feet in width. The ore is of the "mispickle" variety (combination of iron, sulphur and arsenic), carrying gold and silver. Assays of this ore returned $30 gold and $15.30 silver per ton. A tunnel, begun at a proper point, could tap this vein at a great depth. The American Eagle is owned by E. P. Gassman, of Cle-Elum, and Harry Hodges, of Ellensburgh.

THE BOSS MINE.

This property, owned by Col. C. Bell, is also located on Eagle mountain, and on the same vein as is the American Eagle. The ore is of a similar character, and returns have shown $25 in gold and twenty-two ounces of silver per ton. Development of the property is progressing.

MAMMOTH MOUNTAIN,

A lofty peak rising east of Eagle mountain, was first prospected several years ago. Results then did not seem to warrant the expenditure always incident to extensive operation. But they since have. On this mountain are several fine prospects undergoing different measures of development. One of these is the Mammoth mine, owned by James Greeve, of Cle-Elum, and Philip Stanton, of North Yakima. The vein in size is about six feet, carrying gold, silver and copper. A cross-cut has been driven a distance of seventy-five feet, cutting the ledge about fifty feet from the surface.

THE BRONCHO LOCATION,

On the same mountain and owned by the same parties, has a vein of free milling ore, carrying sulphuretts, which is about four feet in thickness. Assays of ore returned about $40 in gold per ton. About eighty feet of development work has been completed on the property.

Contiguous to Fish lake, and in Stevenson gulch, may be found the Silver Bull claim, where five feet of white quartz, carrying iron pyrites, has been uncovered. The ore is free milling, and returns from samples showed $100 in gold and silver per ton. Work completed comprises four tunnels, one forty, one seventy, one ninety, and one 130 feet in length. The last named will tap the vein at a depth of 400 feet from the surface. The Silver Bull is being operated by James Greeves, E. P. Cassman and August Sassi, all of Cle-Elum, and is a fine prospect.

Messrs. S. S. Hawkins and James Greeves, of Cle-Elum, have a decidedly fine looking group of prospects they have designated the Ida Elmore and Valinia. The ore in each is a free milling gold ore. Returns from samples of this ore showed $58 in gold and $2 in silver per ton. The Ida Elmore vein is ten feet in thickness on the surface. A tunnel is being driven at a distance of about 200 feet below the croppings to tap this latter vein.

THE SILVER BOW LOCATION.

On Hawkins mountain, east of Mammoth mountain, is located the property of Ben Kelly, of Tacoma, and Judge Boyles, of Cle-Elum, known as the Silver Bow location. This property is a copper producer. The thickness of the vein is four feet, and assays of the ore returned $26 in gold and twenty-two and one-half per cent. copper. Development is confined to an incline thirty feet in length, in ore.

THE CLE-ELUM MINE

Is situated west of the Silver Bow, on Hawkins mountain. This property was formerly owned by the Cascade Development Company, of Tacoma. Here a vein four feet in width is to be found, of iron sulphide ore, carrying gold and silver. Assay returns on this ore were $55 in silver, $4 in gold, and eighteen per cent. iron, per ton. Work completed consists of an incline driven through ore a distance of seventy feet. On the dump were about fifty tons of ore. The Cle-Elum is now the property of Messrs. Kelly and Boyles. The Hawk, also owned by these gentlemen, is an extension of this claim.

THE HUCKLEBERRY CLAIM

Is owned by Messrs. Swain and Haight, of Roslyn. The vein is three feet in width, carrying copper ore. Assays from this ore showed $39 in silver, $3 in gold, and twenty-two per cent. copper, per ton. Work of development is being prosecuted, and a tunnel is in process of driving to tap the vein a distance of 200 feet below the croppings.

OTHER PROSPECTS,

Comparatively but little developed, are to be found in the district, those of which are deemed worthy of further development being mentioned below.

Those of Thomas McNulty are called the Eureka, the Robert E. Lee, the Tacoma, the Glover and the Davis claims. These, judging from present appearances, certainly deserve further attention from their owners, as do also the Silver Dump, Silver King, Madeleine, Fortune and Bald Eagle claims. Among the more promising prospects located in the district the past year, may be mentioned the King Solomon, Aurora, Silver Belle No. 2 and Once There locations.

SWAUK MINING DISTRICT.

East of the Cle-Elum mining district lies the mineral division above named. Like the last named, it is a pioneer mineral subdivision of the state. As this chapter is wholly devoted to mineral bearing quartz, I shall refer to the wealth of resource of the Swauk district in that department of this report applying to placers and placer mining, as in this latter regard is the Swauk district best known.

Since the issuance of my first annual report (Mines and Minerals of Washington, 1890), a discovery of mineral bearing quartz has been made that certainly warrants an inspection by mining men of this field, with a view toward the definite ascertainment whether Swauk is not valuable equally as a mineral bearing quartz district, as it has certainly been demonstrated a rich and seemingly ever living placer field. The discovery referred to has been located, the finder designating his claim as the

GOLD LEAF MINE.

This property is situated on the east fork of Williams creek, a tributary of the Swauk river. The owner is G. W. Seaton, of Teanaway. It is a free milling gold quartz proposition, of seemingly great merit. The vein approximates six feet in thickness. Milling returns on six tons of ore were at the rate of $40 per ton, with a loss of over fifty per cent. of the assay value.

Should research develop the fact that Swauk district is a valuable mineral bearing quartz section, it will certainly be accepted by the miner as a "poor man's paradise." The country is open and the cost of wagon roads can be minimized. Then, too, these need not be long, as railroad communication, both east and west, is closely beside the boundaries of the district.

PESCHASTIN MINING DISTRICT.

The Peschastin district must also wear the laurels of antiquity, if I may use the term, with both the Cle-Elum and Swauk districts. Within its confines the precious metals were known to be existent as early as 1874, when a veteran prospector named C. P. Culver found on Peschastin creek, a stream traversing the district, paying placers. Some fabulously rich placer claims existed, and are now being found in Peschastin district, which I will refer to in the proper place. But, unlike Swauk district, the Peschastin district by practical demonstration has been judged valuable as well for its mineral bearing quartz possessions as for its placers. The discovery of this latter variety of mineral dates back almost as far as the discovery of placer gold, and to the credit of Mr. Culver must also be placed the finding of mineral bearing quartz. This event occurred in 1878, and in the same year the first stamp mill erected in Washington was built by E. W. Lockwood, esq., now of Okanogan county.

Among the principal mineral bearing quartz claims in the district are the members of a group of mines owned by the Culver Mining Company, of which Mr. J. L. Warner is the general manager. These mines, I understand, are to receive a vigorous development, and to the end that the value of their product may be ascertained, a twenty-stamp mill of modern design has been erected and is ready for operation. This mill will prove at once a boon to a badly neglected district, and whether the ores of the famed Peschastin are as valuable as were and are its placers. The gentlemen who have taken such a commendable interest in solving the problems I have named are Hon. Thomas Burke and Thomas Johnson, esq., of Seattle, and ex-Collector Bash, of Port Townsend.

Aside from the properties controlled by these gentlemen are the following named prospects, which, were they given practical and experienced handling, might aid in the work of demonstrating old Peschastin worthy the attention of operators and capitalists: Humming Bird, Pole Pick, Golden and Phœnix.

MINES OF WESTERN WASHINGTON.

CASCADE MINING DISTRICT.

This is one of the oldest of the mining districts in northern Western Washington, dating its formal organization back to September, 1889. Messrs. G. L. Rouse, J. C. Rouse, Harry Frank and others founded it as the result of important mineral discoveries made by them early in the year mentioned. The district embraces all the area comprising the northwestern portion of Skagit and the southeastern portion of Whatcom counties, and is one of the largest in the state.

TOPOGRAPHICAL.

The general topographical aspect of Cascade mining district presents a rugged mountainous appearance; the area drained by several streams, the principal one being the north fork of the Cascade river, which finds its source in a glacier located in the northwestern corner of the district; and the south fork of the same stream which finds its source near the center of the district. The Cascade river, or main waterway of the district, flows from the northern portion of the district to a confluence with the Skagit river at a point known as Marble mountain.

GEOLOGICAL.

The country rock of the district is found to have a general trend toward the northeast and southwest, dipping from sixty-five to seventy degrees to the south. This applies only to that portion of the district lying on the northeastern side of the Cascade range which practically forms the axis of this district. On the southwestern side of this axis the dip of the country rock is found to vary from sixty-five to seventy degrees in a northerly direction.

The character of the country rock is mainly syenitic granite, mica schists, porphyry, highly metamorphic slates, and quartzites. This country rock is traversed by mineral bearing veins of ore, the general trend of which are found in conformity with that of the country rock, as sectioned hereinbefore. Croppings of these veins are to be found on a similar level to the water courses of the district, and as well, at altitudes on the mountain ranges as high as 10,000 feet above the level of the sea.

ORES OF THE DISTRICT.

The croppings of the mineral bearing veins in the district are found, when not covered with an iron cap-rock, to be composed of galena and iron pyrites. Croppings are found to be from six to twenty feet in width. Galena, iron pyrites, marguisite, brittle silver and copper sulphides comprise the ores found in the district.

MINES OF THE DISTRICT.

The principal mines of the district are found on the northeastern flanks of the Cascade range. It is in this portion of the district that the Boston mine is located.

This claim, the property of J. C. Rouse and G. L. Rouse, was located on the 5th of September, 1889. The location is on the north fork of the Cascade river. The vein is from six to twenty feet wide, and has been traced thus far a distance of 12,000 feet. The trend is northeast and southwest, dipping 70 degrees to the south. It is a contact vein, with a quartzite hanging, and a porphyry foot, wall. The ore is a quartz, carrying iron sulphide, galena, gold and silver. The vein has been stripped a distance of 1,500 feet. Further development comprises a tunnel thirty feet in length on the vein, a shaft forty-five feet deep on the vein, and two drifts at different places run in on the vein sixty and thirty feet respectively.

ANALYSIS—BOSTON LODE.

Assays of samples from tunnel, thirty-foot drift, sixty-foot drift and croppings returned:

Minerals.	*Tunnel.*	*30-foot drift.*	*60-foot drift.*	*Croppings.*	
Gold, ounces	.10	.10	.10	.10	Per ton. 2,000 lbs.
Silver, ounces	48.00	54.00	62.00	134.00	
Lead, per cent	42.00	44.00	48.00	59.00	

THE HARTFORD LOCATION

Is the first eastern extension of the Boston mine. George Sanger, of Mt. Vernon, owns this property. In width, character and value of ore the vein is about the same as the Boston mine.

The Sierra Grande location is the eastern extension (No. 2) of the Boston mine. It is the property of Grant, Ferguson, Getchell & Co. Only the requisite amount of assessment work has been completed on this claim.

The Ontario claim, owned by the same parties, is the eastern extension (No. 3) of the Boston. Assessment work represents the development thus far accorded it.

The Chicago claim is the first extension westward of the Boston claim, and is the property of Messrs. Landry and Brennan.

Western extensions Nos. 2, 3 and 4 of the Boston are the Cincinnati, New York and Buffalo locations, on which assessment work also represents the measure of development. These claims are owned by Messrs. Gilbert, Landry, Landers & Co.

A CROSS VEIN

Of the Boston vein, running east and west, is called the Alta vein. On this vein four locations have been made. As regards width of vein and character of ore, this vein is similar to the Boston vein, or mother lode.

Messrs. Hainsworth & Co., of Oakland, Cal., and Seattle, own the Alta claim. Here a tunnel seventy feet in width has been driven in on the vein. This comprises the development work completed on the claim.

The Montreal, owned by Messrs. Gilbert, Landry, Landers & Co., is eastern extension No. 1 of the Alta vein, and the Cerrico and Helena Butte, owned by Ferguson & Co., are extensions Nos. 2 and 3 of that vein. On all three only assessment work has been done. These locations comprise all thus far made on this vein.

Paralleling the Boston, or mother lode, is what is known as the West Seattle vein, of the same general width, and carrying about the same character of ore as the former. On this vein have been located the Mitus, Diamond and West Seattle claims, on which only assessment work has been completed.

A cross vein of the West Seattle lode is called the Kildare vein, which in all essential particulars resembles the West Seattle vein.

Here are located the Kildare and Harrison claims, comparatively new prospects.

East of the West Seattle is the Soldier Boy vein, trending northeast and southwest, and possessed of the same general characteristics as the Boston and West Seattle veins. The Soldier Boy and six new locations are the property of Messrs. Getchell & Co., of Seattle.

Crossing the river and on the opposite side of the valley from the Boston vein is the Johannesburg claim, owned by Messrs. F. W. Dunn & Company. The vein here is five feet in width, trending northeast and southwest, dipping toward the south. On the property a tunnel has been driven in on the vein a distance of fifty feet.

The Baltimore prospect, owned by Gussey & Dennis, lies north of the Johannesburg, and only assessment work has been done upon it.

ON THE SOUTH FORK

Of the Cascade river, on the mother lode, is located the Cascade mine, owned by McIntosh & Daly. The vein is five feet in width, carrying gold, silver and lead. A tunnel 100 feet in length has been driven in on the vein, and about twenty tons of ore are on the dump. This is a promising looking property. Assay returns of samples of ore from tunnel and dump returned:

Minerals.	*Tunnel.*	*Dump.*	
Gold, ounces	tr.	tr.	Per ton. 2,000 lbs.
Silver, ounces	48	51	
Lead, per cent	32	37	

The Ironclad and Silver Butte properties show croppings of ten feet in width of ore similar to that in the Cascade claim, and are owned by the same parties.

ON THE MIDDLE FORK

Of the Cascade river, two miles and a half from Mineral Park post-office, and thirteen miles from Marble mountain, is located the Epoch mine. Here a ledge of solid galena ore has been found, three feet in width. A tunnel has been run in on this vein, and there are several tons of ore on the dump. Assay returns of samples of this ore taken from croppings, breast of tunnel and dump were:

Minerals.	*Croppings.*	*Breast of Tunnel.*	*Dump.*	
Gold, ounces	tr.	tr.	tr.	Per ton, 2,000 lbs.
Silver, ounces	102	39	41	
Lead, per cent	45	42	38	

Adjacent to the Epoch is a prospect on which only assessment work has been done, owned by Harrison, Marshall & Co. Other prospects on the west side of the range, which I have stated is the axis of the district, are the El Dorado (Millett, McKay & Co.), King Solomon (Everett & Co.), Look-out, Prospector's Friend and Fargo-Union.

ON THE EASTERN SIDE

Of the axis are located several good looking claims. In Horseshoe basin is located the Quien Sabe lode. The vein is sixty feet in width, of quartz and galena, carrying gold, silver and lead. A drift has been run in on the vein for a distance of 100 feet. It is a contact vein, lying between quartzite and porphyry, its trend being northeast and southwest, with a dip toward the north. There are about fifty tons of ore on the dump. The claim is that of J. C. Rouse, of Woolley, Harry Frank, of Tacoma, and Adolph Behring, of Birdsview, Skagit county. Assay returns from croppings, drift and dump were:

Minerals.	*Croppings.*	*Drift.*	*Dump.*	
Gold, ounces	tr.	.10	.10	Per ton, 2,000 lbs.
Silver, ounces	204	103.00	108.00	
Lead, per cent	48	46.00	54.00	

An extension of this property is the Quien Sabe No. 2, owned by the same parties. But little work has been done on this claim. Ferguson & Co. own extensions No. 1 and No. 2 of the Quien Sabe No. 2 which they have called the Black Warrior and Blue Devil.

Near the Quien Sabe is the Doubtful mine, where the vein is fifteen feet in width, with a pay streak of twelve inches of galena ore. This is a concentrating proposition. Two tunnels have been run in on the vein; one thirty, the other 100 feet in length. About twenty tons of ore are on the dump. Assay returns of samples from drifts Nos. 1 and 2, and from croppings and dump were:

Minerals.	*Croppings.*	*Drift No. 1.*	*Drift No. 2.*	*Dump.*	
Gold, ounces	tr.	tr.	tr.	tr.	Per ton, 2,000 lbs.
Silver, ounces	100	95	97	87.50	
Lead, per cent	60	54	62	59.00	

Also contiguous to the Quien Sabe is the Galena claim, owned by Rouse & Co. The vein is seven feet in width, of carbonate ore, carrying gold, silver and lead. About $600 has been expended thus far in the development of the property: Ten tons of ore are awaiting shipment. Assays of samples from croppings and dump returned:

Minerals.	*Croppings.*	*Dump.*	
Gold, ounces	tr.	tr.	Per ton, 2,000 lbs.
Silver, ounces	72	90	
Lead, per cent	54	61	

The White Cap, Comet, Lady of the Lake and Water Fall claims in the neighborhood are fine looking prospects belonging to Kingman & Pearshall.

THE FRANKLIN LOCATION

Is one of the most promising in the district. It is located north of the Quien Sabe mine, and is owned by Harry Frank, of Tacoma, and Messrs. Rouse & Rouse, of Woolley, Wash.

A vein has here been exposed which is four feet in width, carrying gold, silver and lead. A small measure of development work has been completed. Returns of assays of samples of the ore taken from the croppings and dump were:

Minerals.	*Croppings.*	*Dump.*	
Gold, ounces	tr.	tr.	Per ton, 2,000 lbs.
Silver, ounces	100	89	
Lead, per cent	54	42	

Messrs. Behring, Speckler and Gates own an extension of the Franklin called the Adolph. Only assessment work has been completed.

Messrs. Campbell, Hill & Co. have been operating extensively in the district the past year. This firm owns the following likely looking prospects, the figures opposite each named representing the width of the vein. The ore in each is a carbonate and galena, carrying gold, silver and lead:

The Ohio, 3 feet; the Indiana, 2½ feet; the Maryland, 4 feet; the Great Republic, 3 feet; the Michigan, 2 feet; the Big Chief, 3 feet, and the Roscoe Conkling, 2 feet.

ACCESSIBILITY.

While probably, from a topographical standpoint, embracing an area as rough as that of Monte Cristo district, Cascade mining district is easily accessible, and once arrived within its boundaries, the visitor will find numerous trails leading about it. The principal method of ingress and egress is via steamer from either of the principal commercial centers on the Sound shore line to Marble Mountain, a point distant but twenty miles from the mines, thence by pack train to the mineral belts. Communication with the district may be made also by rail to Hamilton, sixty miles distant from the mines, which may be reached by good wagon roads and pack trails.

MONTE CRISTO'S WEALTH.

A district, or rather what might be more appropriately styled a mineral division of Northwestern Washington, of comparatively recent formation, and yet one that even at this pioneer stage of its development at the hands of hardy prospectors is being daily brought into greater prominence, and is attracting the attention of capitol, is Monte Cristo district. This district, to my mind, contains the natural extension of the great mineralogical zone in the part of Washington designated, a part of which and the measure of research into the formation and value of which I have noted under the caption of the "Silver Creek mining district, Snohomish county."

Away up in the Cascade range of mountains, nestling as best it may among the peaks and serried bluffs and glaciers of a region as romantic as it seems to be impregnable to the casual observer, lies, in my opinion, a district calculated at no distant day to win by its merits the laurels of pronounced distinction as one of the very best seats of this state's mineral wealth.

But to more particularly describe Monte Cristo mining district, let me say that it forms a portion of the eastern boundary of Snohomish county; its own boundaries traversing a series of spurs of the Cascade range. In fact, its general topographical aspect indicates its territorial composition, in the main, to be nothing more nor less than these giant arms of the main range, each in turn, and with, to me, a remarkable uniformity as to line and extent, jutting out over the country to the westward for a number of miles.

That the Monte Cristo mining district is topographically of the roughest possible character will go without gainsaying to any one who has ever paid a visit to its lofty confines, climbed the rugged mountain peaks abounding—in fact, absorbing—its territorial area, or sought for treasure trove in and among its precipitous and grand arroyos and ravines, or attempted survey of its surroundings. Still, incontrovertible evidences to the venturesome gold seeker that the all-coveted mineral lay hidden in its equally well hidden recesses impelled him to overcome all obstacles, to penetrate its fastnesses, and then, finding his anticipations realized, to fly back to civilization, herald the good tidings, and, with luxurious ease compared with what he had undergone in seeking his treasure, guide the army that follows always in the wake of the pioneer to his "find."

With the growth of the district, and the increased interest manifested in it by miners and mining men, has naturally followed a research to ways and means to make it not only more easily reached, searched and returned from, but placed in comparatively close proximity to the several commercial settlements which may, strange as it may seem, be said to prosper almost at its gateways. As a result of judicious and combined effort in this direction, Monte Cristo is now as available to the traveler as Silver Creek or any of the several mining districts dotting the western slope of the Cascades.

Taking an air line to the eastward, this district may be reached from Seattle by traversing a distance of only fifty miles. An air line followed from the northward and traversed forty miles brings about the same result.

Three comparatively easy methods of ingress and egress are offered the traveler to and from the district. First, he may leave a Puget Sound commercial center, journey to the city of Snohomish, and thence by easy stages proceed up those two tributaries of the Snohomish river (Skykomish river and Silver creek), to the mines of Monte Cristo; second, he may travel east from the prosperous and hospitable little agricultural town of Arlington, in the fertile valley of the Stillaguamish, to the mines; and lastly, but far from least, because it is a popular method of reaching Monte Cristo, he may start from the city of Sedro up the valley of the Skagit to the Sauk, and be at the very seat of operations in short order.

HISTORICAL.

Monte Cristo mining district, there can be no doubt, owes its existence to the discovery of mineral wealth on the Sauk river, a magnificent mountain stream, at the head of which it is to be found, and which stream, flowing in a southwesterly direction, empties into the Skagit river at the head of navigation of the latter stream; and at a point where a lively little hamlet yclept Sauk City may be found.

Mineral wealth on the Sauk first attracted the attention of prospectors in August, 1889, and "an excitement" speedily followed the location of the first mining claim in Monte Cristo district. It was christened the '76, and its locators were those veteran prospectors, Messrs. Pearsall and Peabody. The district proper of Monte Cristo was formally organized by Messrs. Willman, Pearsall, Peabody and others in the same year. Since that time the possibilities of the district have attracted the attention of a large number of miners, and as a result, infantile as it is in years, Monte Cristo has attracted a surprisingly large share of attention.

GEOLOGICAL.

From a geological standpoint, the district offers just as splendid a field of study and research to the practical geologist as, in my opinion, does the Silver creek region. Contemplation of its formation, study of glacial characteristics there abounding, and an insight into nature's choice of formation for this aerial region, will in themselves call upon the student for effort in his researches his surroundings impel him to gladly volunteer in full measure.

The country rock of the district largely comprises highly metamorphosed slates, these being curiously twisted and tilted to the high angle of, approximately, sixty degrees to the west, and their exposure occurs in what is termed the "'76 basin," on the south fork of the Sauk river. This same group of slates, as regards Glacial basin, to the east of '76 basin, is to be found broken off by glacial action; in fact, well nigh completely removed, and at this point may be found exposed abrupt granite peaks, cerrated with numerous porphyritic dikes, these latter trending northeast and southwest, and being highly mineralized and constituting in popular vernacular, "the veins of the district." Decay occurs with these dikes or veins much speedier than with the slates, this being due to their upheaval. Taking into consideration the established fact that the ore accompanies these veins, and the further fact that they are evidently upheavals, and that the country rock at right angles is bisected by them, both granite and slate formations, the deduction is drawn that the veins in this region are true fissure veins.

CHARACTER OF MINERAL.

The character of the ore found in the district is a sulphide. Among the slates the ore is mainly a galena, carrying silver, while in the granite of the great glacial basin iron sulphide, carrying gold, abounds. In other portions of the district, notably further down the river, galena and iron sulphide, carrying some copper, are to be found in paying quantities, and I am confident that a careful and systematically waged quest will bring to light new finds in comparatively unknown portions of the district, basing my opinion in this regard upon my knowledge of its geology.

CLIMATE.

The climate of the district I might best describe as follows: Cold and bracing during the early spring months, balmy and delightful throughout the summer, cold, raw and appetite inspiring during autumn, and exceedingly cold in winter. At that altitude, however, I must say I was surprised at the equability of each member of Monte Cristo's series of climatic conditions. These naturally vary. A winter of comparative climatic peace may be followed by one of war among the elements, decidedly uncomfortable to the inhabitant. Taken as a whole, the climate of Monte Cristo may be said to be far superior to that of the mining regions of Montana and Colorado in spring and summer, and but a poor competitor of the same during the other seasons of the year.

TIMBER.

High as the altitude of Monte Cristo is above the sea level, its possession in the shape of timber is one of its most valuable features. The growth is as a rule strong, healthful, useful, and therefore, valuable. Fir, mountain pine, spruce, cedar, maple, ash and cottonwood are to be found in abundance in and about the district boundaries. Timber valuable alike for fuel, building and mining purposes I found to be abundant on the Sauk river, and at the site of one of the mines in the district the timber growth was of a splendid character.

WATER.

I need hardly report that in Monte Cristo water is a continual drug upon the local market. Mountain streams of greater or less extent abound all over its surface, and springs and lakes abound. The mining interest will never suffer from lack of the fluid, nor will the inhabitant who may have resided there from his natal to his dying day.

MONTE CRISTO MINES AND MINERALS.

So young is Monte Cristo, that, although active operations, and on no small scale, have been started on development work, those having them in hand had actually no time in which to make the showing that a camp at such age as Silver Creek, for instance, can exhibit. Still, with the facilities at hand in a new district, and the time allotted its founders and inhabitants, such a showing of its resources has already been made as to establish in the minds of the Monte Cristo residents the firm belief in the richness and permanency of its mineral possessions. Development work on all claims, as thus far conducted toward the ascertainment of their character and value, demonstrates to me that complete, systematic and intelligently directed work is warranted on the part of their owners, and I have no hesitancy in suggesting this conduct on their part.

From a practical miner's standpoint, the district is certainly a paradise, as the cost both of preliminary work and actual operation may easily be reduced to a minimum by the exercise of even ordinary economy. Tunnel sites abound where ledges exist. Waterways are at hand at each of numberless mill sites; fuel is at the prospector's door, or the site of a mill, hoisting plant or boarding house. In short, let me say for Monte Cristo mining district that, in so far as natural surface advantages and opportunities afforded the miner on top are concerned, it has no equal in Washington.

SOME FINDS OF PROMISE.

The Pride of the Mountains is the title of a find of promise in Monte Cristo district. This property is located on the east side of the great glacial basin, at an altitude above sea level of 4,050 feet. Development work has already shown an eight-foot vein, the ore being quartz and galena. An assay made from a sample taken across this vein returned $10.60 in silver and $9.71 in gold; another assay from the cropping of the vein returning $79.60 in silver and gold and 58.06 per cent. in lead per ton (2,000 pounds). Development work comprises a tunnel run a distance of fifty feet on the vein, and an open cut, fifty feet from the mouth of the tunnel, driven twenty feet on ore. When I saw the property, about fifty tons of ore lay on the dump. Samples of ore from this dump assayed $26.40 in silver and $8 in gold per ton (2,000 pounds).

THE EASTERN EXTENSION

Of the Pride of the Mountains is called the '89, and its altitude is 4,600 feet above the sea. Here ten feet of ledge matter has been unearthed. A sample from the vein assayed $30 in gold and silver to the ton of 2,000 pounds.

Between the Pride of the Mountains and the '89 is to be found the White Cloud, on the same vein; and to the westward of the Pride of the Mountains lies the Pride of the Woods, at an altitude of 4,230 feet. The ore here found is concentrating, carrying but little silver, no lead, and averaging one ounce of gold to the ton of 2,000 pounds.

A western extension of the Pride of the Woods is found in a new claim called the Clara.

To the southward of the Pride of the Mountaius vein may be found the I. X. L. vein, and located with the I. X. L. upon it are the Mystery, altitude, 4,280 feet, and the Side Line. At the Mystery mine, from the face of a drift, at a depth of about ten feet, the vein was shown to be six feet in width, holding ore of a concentrating character. An assay of this ore returned $9.20 in silver and $12.40 in gold per ton (2,000 pounds).

On the western side of '76 basin are the Glacier, Uncle Sam and Emma Moore claims. The latter has a twenty-foot outcrop of ore, assays from which returned $40 in gold and silver and fifteen per cent. lead.

South of the Emma Moore vein, and running east and west across '76 basin, is the '76 Mammoth vein, the altitude of the tunnel being 5,000 feet, on which are located the following named claims: '74, '75, with tunnel 4,100 feet above sea level, '76, Ranger and Pinnacle. The '76 claim has a vein twenty-eight feet wide. An average sample of the ore of this vein returned $30 silver, $22.73 gold, and twenty-six per cent. of lead, per ton (2,000 pounds.)

THE RAINEY VEIN.

At the confluence of the North and South forks of the Sauk river is located the Rainey vein — on the north side of the junction and a short distance above it. This vein, carrying a heavy deposit of iron sulphide ore, has a thickness of about fifteen feet. The vein has been cross-cut a distance of about thirty feet. Average assay returns from this ore were $15 in silver and $30 in gold per ton (2,000 pounds); from the four-foot pay streak, $60 in gold.

The eastern extension of the Rainey vein is called the Phœnix; but little development has as yet been accorded it.

All the above described claims (from the Pride of the Mountains to the Phœnix) are the properties of Hon. Thomas Ewing, Judge Bond, the Willman Bros., Joseph Pearsall *et al.*, of Seattle. This company, in connection with the citizens of Skagit county, completed an excellent wagon road a distance of fifty miles from Sauk City, head of navigation on the Skagit river, to the mines, and their disbursements in this and other matters, in connection with their interests in Monte Cristo, must approximate $65,000. The gentlemen are conducting operations in a most systematic and commendable manner, and at their center of operations, below the Rainey vein, is a camp that would be a credit to a much older district.

I saw many likely looking prospects, in fact, an army of them, and will here take occasion to mention the Orphan Boy, Central Fraction, Climax, Washington, Sidney, Tuscolo, West Seattle, Russia, Laborum, Saturday, Stratas, Homestake, Nibus, Keystone and Hydra, all showing commendable development, and their holders exhibiting unbounded confidence in their future greatness.

SILVER CREEK MINING DISTRICT.

One of the most important mining districts in Washington, and one undergoing as great a measure of practical and systematic development as any district in the state, is what is designated the Silver Creek mining district.

In my previous report (Mines and Minerals of Washington, 1890) I took occasion to refer to this district as one of promise, and well worthy practical and intelligent investigation. I am pleased to be able to here chronicle the fact that a full measure of each has been accorded the district; and at the hands of men well calculated by experience, both of a practical and scientific character, to conduct this important work.

As a result of quite a complete measure of preliminary development of its mineral resources, Silver Creek district has jumped to a position in the front rank of the several valuable mineral localities of the state — and their name is legion, as I have demonstrated; and has attained to this rank, practically speaking, in the short space of a twelve-month. The attainment of this high rank by the district, those there interested and the public generally should remember, is wholly due, first, to the perseverance, industry, and confidence in the future of the district, of the pioneers in the work of making known its mineral wealth; secondly, to a remarkably businesslike persuasion in bringing capital into the district, thereby establishing firmly its complete and correct development, which has characterized both pioneer and newcomer there. Capital, let me say here, is as essential a requisite to the attainment of a mineral belt to commercial prominence as air is to the maintenance of life; and, recognizing this truism, the miners and operators of this district have gone systematically to work; illustrated practically and with evident thoroughness the value of their possessions; conservatively advertised them abroad, and secured the essential I have referred to above; and with it sure promise of far greater reward in the near future than has repaid their efforts in the past. Combined business ability and the ability to conduct what I might term practical manual labor, those intending to or those actively engaged in the work of developing our state's mineral wealth must understand, are essentially necessary and requisite to complete to unqualified success the pursuit of that most fascinating of vocations, mining.

DESCRIPTIVE.

I quote from the report of this office issued January 1, 1891, *i. e.*, "Mines and Minerals of Washington, 1890:"

"This Snohomish county mineral division is located in the eastern part of the county, and takes its name from a beautiful stream of water of that title, which flows westerly from the summit of the Cascade mountains through a canyon a distance of eight miles into the north fork of the Skykomish river. The main creek is fed by numerous small creeks running from the silver belt through deep canyons. From the source of Silver creek to its mouth is one continuous mineral belt. The country rock of the Silver Creek district is porphyry, granite, diorite slates; well defined and unbroken ledges of galena (a combination of lead, gold, sulphur, silver and iron) can be traced for great distances on the surface. In my official capacity I visited Silver Creek district. I inspected over thirty ledges. Some of the galena ore I found carried arsenical iron, and ran very well in gold. Among the principal mines inspected were the Blue Bird and Vandalia properties."

Additionally in this regard, let me note in this, the second report of this office, facts garnered by subsequent visits to this district, and results of examination and extended investigation conducted therein.

GEOGRAPHICAL.

From a geographical standpoint the district might best be described as follows:

Silver Creek mining district is located in the eastern corner of the county of Snohomish, in northern Western Washington. The district is bounded to the northward by the Monte Cristo mining district; to the eastward by the Cascade mountains; to the southward by Snoqualmie district; to the westward by a rich agricultural district, famed throughout the state. For a comparatively new district it is most advantageously located as regards natural and artificial methods of transportation, terminating from its boundaries at recognized commercial centers of trade.

Situated in an air line about thirty miles from the busy and rapidly growing mining, commercial and manufacturing city of Snohomish, county seat of the county, Silver Creek district may be easily reached from this, its *entrepot*, by good roads which will guide the visitor into the very heart of the district.

About sixty miles from the center of the district is situated, to the southwest, the city of Seattle, one of the state's metropolii and one of the most important of northwestern municipalities. As described, the district is indeed most advantageously situated; and I predict will soon become a potent factor in the upbuilding of the communities of greater or less degree, to which it will be a tributarial source of wealth and progress by its very geographical relation to them.

TOPOGRAPHICAL.

The topographical features of this district at once strike the stranger within its boundaries as a most romantic and really enchanting combination of the rugged and sublime in nature—a characteristic noticeable, to my mind, throughout Northwestern Washington, where the fertile valley is shaded by the towering mountain, the chill snow breezes softened by the zephyrs of the valleys and dales. The district proper may be described as a succession of serried peaks, rolling onward and upward to the Cascade mountains; these peaks here and there interspersed with charming, and, from an agricultural standpoint, exceedingly fertile valleys of larger or smaller dimensions. Enough agricultural territory is available in the district to satisfy any demand made upon it by consumers for some time to come; and any anxiety in this regard may be quelled because of the known richness of the agricultural lands of the county in which it is situated.

WATER AND WATER WAYS.

Especially fortunate is Silver Creek district in the possession of an absolutely unlimited supply of water, of a character unexcelled. Prominent among the streams traversing its surface I might mention the Skykomish river, an officially charted stream, and a navigable water way for a distance within twenty miles of the mines from its confluence with the waters of the Snohomish river. The Skykomish rises in an exceedingly rugged portion of the Cascade mountains, contiguous to the eastern boundary of the district, flows thence in a westerly way to the Snohomish river, its waters, aided by those of the latter navigable stream, being carried into Puget Sound at a point near the city of Snohomish.

Silver creek, whence the district takes its name, is a rapid mountain stream, taking its source in the mountains in the northeastern portion of the district, flowing thence in a southwesterly direction to a confluence with the Skykomish river. Silver creek is not navigable, but its value, turned to the uses of manufacture and those commercial relations always incident to mining, cannot be overestimated. Among other mountain streams, the waters of which will soon be turned to advantage and profit, I might name Troublesome, Cascade, Lost Trail and other creeks. A chart of the water ways and general water supply of this district would delineate to the observer a territory more than usually endowed with this valuable commodity. Ever living springs abound, and nature is not loth to supply both rain and snow to the country with a bountiful hand during the winter season.

CLIMATE.

As is characteristic of entire Washington, the climate of Silver Creek district is most salubrious. Naturally, its high altitude causes snow to be a frequent, but usually a friendly visitor in winter. The climate of the spring and summer months is simply delightful; and frequent precipitations of rain make clear, wholesome and invigorating the general climatic condition of the district.

TIMBER.

A wealth of magnificent timber was one of the gifts of nature to Silver Creek. All the giant species so famous in the mountains of Washington are there to be found. Fir, tamarack, cedar, are there in abundance, ready to play their important *role* in the drama of the district's mineral development. The attractive flora of the Cascades is not missed there, and the general *ensemble* of the timber panorama is at once highly attractive and grand, and it should go without saying this timber is destined to play a most important part in the upbuilding and perpetuation of the mineral belt.

HISTORICAL.

Concluding what may be termed prefatory remarks concerning this district, a brief record of historical data in this connection might now be in order.

Eight mineral locations noted the advent of Silver Creek in the ranks of the state's mining districts. These were filed in the office of the auditor of Snohomish county in midsummer, 1871, or twenty years ago. Aside from the filings mentioned, which were those of

George White and Hill Tyler, nothing was done that year. In fact, over three years elapsed before operations in this district were commenced. In 1874, L. T. Ireland and John Richards delivered to a chemist for assay purposes, samples of ore taken from what they called the Albert vein, but now prominently known as the National mine. The result of these assays warranted a small army of prospectors to enter the district. As a result, innumerable locations were filed, and that irrepressible excitement-follower, the townsite expert, made his appearance. The platters christened the first embryo municipality in the district Silver City.

In that early day Silver Creek district was isolated from the world, and journeys to and from it attended with all the discomforts, even dangers, and great expense of pioneer time in any region. Remote from civilization, although it was to be found at its very door, so to speak, the district had but little charm to the innumerable "tenderfeet" who heard of it and rushed thither. Their stay was not prolonged, and late in 1874 Silver City, with the exception of the indomitable Ireland, John Cochran and George White, was minus inhabitants. For six years this trio of gold seekers "staid with" the city. In 1880 a primitive mill of the Spanish arastre species was erected by a Snohomish company, and the project was looked upon as indeed a daring one. Meanwhile several new leads were found, and in this arastre were given tests. The ore, being too base to suit this crude process of treatment, the Snohomish company's venture proved a failure. This caused a second abandonment of the city and district, and the locations lapsed. It was not until 1888 that the district next came into prominence, although Ireland, Cochran and White kept steadily at their arduous labors as prospectors. In the year 1881 their party was reinforced by the arrival of E. H. Hubbart and Lon Lowe, old miners.

The old camp of Silver City gave way soon afterward to the present Mineral City, E. H. Hubbart and others constituting the new townsite company. Ever since Silver Creek mining district has been prominent, for two years to a greater or less degree, but now ranking one of the most prominent mineral divisions of the state.

GEOLOGICAL AND MINERAL.

Silver Creek mining district offers a splendid field for investigation to the observant geologist. It is bounded on the east by a belt of true granite, extending along the summit of the Cascade mountains north and south and cropping at the head of "Molly-be-dam" gulch. To the westward this formation cannot be traced. I examined in this district a strong mineral belt, with several veins. The veins had, approximately, a uniform strike, it being a few degrees north of west, their dip being almost vertical. The veins are apparently strong, and are well defined.

A vein of blue diorite, approximating twenty feet in width, and apparently forming the center of the axis of this mineral belt, extends from what is known as National mountain to "Molly-be-dam" gulch, to the southward. All veins in this belt are of porphyritic quartz formation, with clean cut walls, talcose lined. In close proximity to the dioritic axis mentioned, the veins apparently hold a greater quantity of galena than do those further removed from the diorite.

Subterranean disturbances seem, at the mouth of Silver creek, to have twisted and entangled the massive beds of granite gneiss, allowing the seams to become receptacles of a volcanic quartz or porphyry. Investigations thus far carried on have been of a character by which nothing authoritative concerning these veins may at this early period be chronicled.

MINES OF THE DISTRICT.

In the first annual report of this office (Mines and Minerals of Washington, 1890) considerable attention was paid the measure of development under way of the then furthest advanced of the prospects. Since the issuance of that report several properties therein referred to have advanced without the category of "prospects," to full-fledged mines.

The active development of every property deemed worthy of it, has been waged with unceasing vigor, and I am glad to note with gratifying and confidence inspiring results.

The ores of Silver Creek district in the main are galena, silver and iron; of the latter metal a sufficient percentage is generally found to make a good smelting ore. Among other varieties of ores secured by me were specimens of tetrahedrite (gray copper), cerrusite (lead carbonate), anglesite (sulph. of lead), minium (ox. lead), several specimens of green, blue and purple copper of great beauty, and some pyrite.

THE VANDALIA MINE.

A property ranking high in the district is that noted above. This mine is owned and operated by a Seattle syndicate of gentlemen of wealth, including Leigh S. J. Hunt, esq., F. L. Leslie, esq., and E. C. Blewett, esq.

The mine is located near the southern extremity of the mineral belt on the west side of Silver creek, to the southward of Mineral City about two miles.

The ore is a carbonate and sulphate of lead. The strike of the vein is northwest and southeast, and it is almost vertical.

Development work, on the occasion of my visit to the property, comprised two cross-cuts and two added levels, aggregating 600 feet of work.

In size the Vandalia vein will average five to twelve feet, with a pay streak of about four feet.

Assays from this property showed $300 silver, $50 gold, and forty per cent. lead, per ton of 2,000 pounds. On the dump of the mine is to be seen some fine looking ore.

THE BLUE BIRD VEIN

Parallels that of the Vandalia, being situated due east from the latter. It is the property of the same gentlemen controlling the Vandalia, and the ore is similar. Development work, on the occasion of my last visit, was being vigorously pushed, and the project of tapping both veins at a depth of 600 to 1,000 feet, with a cross-cut being driven to facilitate the output, was being rapidly carried ahead with the aid of three shifts of men, but since, on account of vexatious litigation has been, I learn, discontinued.

THE IDAHO PROPERTY

Is the west extension of the Vandalia, and is owned by Pearsall & Company, of Seattle and Silver Creek. The size of the vein is about seven feet, with a pay streak varying from six inches to one foot. The ore is a galena ore, and assays made of it showed all the way from $20 to $40 in gold and silver. On the dump were about twenty tons of ore.

THE BILLY LEE

Is a mine located on the west side of Silver creek, two miles south of Mineral City. It is the property of the Hon. A. W. Frater,

Senator Vestal, and Judge Denny of the superior court of Snohomish county.

The ore is a galena, assays from which went as high as $200 per ton (2,000 pounds) in gold, silver and lead. The vein is eighteen feet in width, carrying a pay streak of two feet in width. Its strike is northwest and southeast, and is almost perpendicular. Development work of the Lee comprises a tunnel driven in on a stringer a distance of 112 feet from the southeast boundary of the claim. Six hundred feet above the entrance to this tunnel a tunnel has been driven in on ore a distance of twenty feet. On the dump were about twenty tons of ore.

Other mines owned by these gentlemen associated together as the Snohomish, Silver Creek & Port Gardner Mining Company, comprise the Port Gardner, Even Up, Black Diamond, Frater, Helena, and Ingersoll. All I found in greater or less state of development, and all promising. I am informed these properties will be developed as speedily as possible with a view toward their being made actual producers.

THE RUBY KING,

A very fine looking property, I found northwest of the Port Gardner claim, noted above. Its vein parallels the former, and the quality of ore similar. The vein I found to be eight feet in thickness, with a pay streak of eighteen inches. Assays of this ore have shown $100 in gold, silver and lead, per ton (2,000 pounds). The mine is the property of Messrs. J. B. Field and William Flobeck, of Silver Creek, who propose pushing its development vigorously.

Further down toward the mouth of Silver creek I found many promising claims. Among these were the Evergreen, with a vein seven feet in width, carrying silver, copper and gold. C. R. Howard & Co. own the mine.

The Oro Fino, another claim, is owned by Redlin & Peck, also of Silver Creek; has a vein eight feet in thickness of free silver milling ore.

The Michigan Gulch Mining Company, a Seattle corporation, has a vein of five feet nine inches of quartz ore, carrying some gold.

The Emma lode carries a vein of five feet of ore quartz and iron pyrites. The claim is the property of J. Ware, an old mining man. Mr. Ware has driven a tunnel twenty-five feet on the vein, the strike of which is east and west. This is a very fine looking prospect.

The Delta, another fine prospect, is the property of Superior Judge Denny, of Snohomish. Here a vein five feet in thickness, between walls, has been unearthed. The ore is quartz, carrying copper and lead sulphide. The owner is applying for a United States patent for his property, and will then complete extensive development work.

THE CAPLIN-HOLBROOKE GROUP.

This is the name given some fine looking claims situated on the north fork of Silver creek, about two miles north of Mineral City, the home of the superintendent of the group, Col. William Caplin, an old Colorado miner, assayer, chemist and thorough metallurgist. Comprising the group are the Gold Eagle, New Strike, Last Chance, Mt. Beauty, Fortunate Monarch, Silver Queen, Good Grub, Cougar and Two Hams.

The principal development work is that being vigorously and systematically pushed on the Golden Eagle property. The vein, on the surface, crops ten feet in width. A tunnel has been driven a distance of seventy-five feet on the vein, through ore. There are two pay streaks of twelve inches each. Assays of ore from these showed as high as $90 per ton in gold, silver and copper, the Gold Eagle being essentially a gold proposition. It is certainly worthy the development being accorded it. It is the intention to tap the vein with a tunnel, to be driven at a point 500 feet below the present workings. This tunnel should tap the vein at a distance of about 150 feet from its mouth. I look to this group for a great record as coming Silver Creek district producers.

THE NATIONAL MINE

(Described by me in Mines and Minerals of Washington, 1890), is the property of Messrs. E. H. Hubbart, T. D. Brown and Thomas Lockwood, all miners of experience. The National is situated on the south side of Mineral Peak, about three and a quarter miles north of Mineral City, and in the northwestern part of the district. The strike of the vein is east and west, and is almost vertical. It is three and a half feet in thickness, lying between porphyry and granite. Assays of ore from this vein have shown as high as $111, and low as $60, in gold and silver. The altitude of the mine approximates 4,000 feet.

Development work comprises a shaft sunk twenty feet, through ore, at a point 100 feet below the apex of the vein; a cross-cut, fifty-eight feet in length; and two adit levels, one twenty and one fifty feet, each in ore.

THE JASPERSON LODE,

On the east side of Jasper creek, lies to the north of the National mine, its vein paralleling that of the latter. The Jasperson has a fine looking vein, varying in thickness from four to seven feet. The ore is a galena, and the vein carries a pay streak of sixteen inches. Assays from the Jasperson have shown as high as 138 ounces silver, and forty-four per cent. lead, per ton of 2,000 pounds. The lode is the property of Messrs. T. D. Brown, H. Denn and E. H. Hubbart.

THE EAST EXTENSION

Of the National mine is called the Winnie. The character of the ore is similar to that of the National. The size of the vein is eight feet, with a pay streak of four feet. Assays of ore taken from a shaft of the depth of ten feet, showed $56 in gold and silver, per ton. The shaft referred to comprises development work on this claim. The National, Jasperson and Winnie claims above described are each owned by Messrs. Hubbart *et al.* And these gentlemen propose driving a tunnel that would tap the National vein at a depth of 1,500 feet, the Jasperson at 1,700 feet, and the Winnie and several claims owned also by them.

J. T. HUBBART & COMPANY

Also own the Trade Dollar claim, located on Trade Dollar creek, west of Mineral City. The development work has been slight upon this property since my last report. Located at an altitude of 2,600 feet above sea level, the vein I found to be four feet in thickness, carrying iron sulphide. Former assays showed the ore to carry thirty-seven ounces silver, and sixty-three per cent. lead, per ton of 2,000 pounds.

THE MORNING STAR

Is a claim also referred to by me in "Mines and Minerals of Washington, 1890." E. H. Hubbart and John Maxwell still own this fine property. The size of the vein is five feet, with a pay streak of two feet. Its strike is northeast and southwest, with a dip of forty degrees to the westward. Ore from the vein, under assay, showed from $55 to $100 gold, silver and lead per ton (2,000 pounds). The character of the ore is a quartz, carrying lead and copper sulphide. Development to date comprises a tunnel twenty feet long; two cross-cuts, one thirty feet and one ten feet long. Work will be prosecuted the coming winter on the property, and a good showing should be made.

CROWN POINT MINE.

This location is situated on the east side of Silver creek, on what is called Straightup gulch, and is the property of the Crown Point Mining Company; owners also of the Silver Shield, Helena, Hidden Treasure and Crawford claims, and a mill site contiguous to them. The properties are in charge of A. A. Abbey, an experienced superintendent. The ore found in all these claims is a galena, carrying silver and gold.

The Crown Point has a vein five feet in thickness. A mill run from this ore returned 107 ounces silver and forty per cent. lead per ton (2,000 pounds).

THE HIDDEN TREASURE'S

Vein is four feet in thickness. Assays returned $59 silver and twenty per cent. lead per ton (2,000 pounds).

The Helena carries a vein twelve feet in thickness, of which no assay returns are at hand.

The Silver Shield carries a vein eleven feet in thickness. No assay returns of this ore have been received.

The Crawford's showing is a surface one of twenty feet of ore.

On the Crown Point a tunnel thirty feet long has been driven in on the vein, and it is the intention of the owners to continue this tunnel 300 feet further. No development work of any extent has been done this year on the other members of this group, but the coming year great development is promised.

THE MINNEHAHA CLAIM.

This claim is the property of John A. Campbell, esq.; is on the east fork of Silver creek, two miles above Mineral City. The strike of the vein is northeast and southwest, with a dip of seventy-five degrees to the northward. The ore shows a cropping of twelve feet, with a pay streak of twenty-six inches, consisting of quartz and iron pyrites. Assays of this ore returned $40 per ton in gold and silver. An open cut ten feet in length, and a tunnel in process of driving, comprises the development work completed and in process on the property.

THE JUMBO LOCATION

Is situated on the west side of the east fork of Silver creek. A large crop of ore is here shown. The ore is a quartz and pyrites ore, assaying $2.20 in silver, and $21.08 in gold, per ton (2,000 pounds).

A promising mine also is the Arkansas Traveler. It is located on the east side of Silver Creek, as are also the Old Missouri, Marine and Scotch Lassie Jean.

The Silver Slipper claim, on the west side of Silver creek, is a likely looking prospect. Here a vein with a twelve-inch pay streak of silver glance ore has been undergoing development. The Four Brothers, a parallel claim, has a four-foot vein of galena ore, with a pay streak of fourteen inches. These claims, *i. e.*, the Silver Slipper and Four Frothers, are owned by John Dowd, P. J. Field and William Flobeck.

The Gettysburg, located east of the Silver Slipper claim, is another good location. Here a four-and-a-half-foot vein of galena, iron pyrites and quartz has been opened, and which contains a pay streak of fifteen inches. Assays of this ore returned $40 in gold and silver per ton (2,000 pounds). It is the property of Dr. Wright and James W. Maple.

ON TROUBLESOME CREEK.

This creek, a robust stream, takes its course in the pass of the same name, in the northeastern portion of the district. Its course almost parallels that of Silver creek, and it empties into the Skykomish river at a point about three miles from the confluence with that stream of Silver creek.

On this creek are located some of the most likely claims to be found in this district. Prominent among the individuals and associations developing Troublesome creek discoveries, ranks the

GOLCONDA MINING COMPANY.

This company comprises in its *personnel* Messrs. George H. Thomas, J. C. Bonsall and A. Waldroff, and is the owner of a group of properties of promise near the summit of the divide. Among the members of this group is the Pride of the West, a fissure vein in granite from four to twenty feet in width. The vein carries a pay streak of from one to two feet in width. The croppings of this vein are exposed for a distance of 2,500 feet. The ore is a decomposed quartz, carrying galena and iron sulphide. Assays from samples of this ore have returned from $20 to $30 in gold and silver.

The northwest extension of the Pride of the West is the Victoria lode. The croppings are exposed for a distance of nearly 2,000 feet.

This company is now engaged in development work on its original location, the Monte Cristo claim. A tunnel is being driven, and has reached a distance of about forty feet. In the breast is exposed about ten inches of galena on the foot wall and about nine inches on the hanging wall. Assays of Monte Cristo ore have returned fifty ounces silver and forty-two per cent. lead per ton (2,000 pounds). Small shipments of ore from the mines of the company have been sent in for treatment, and encouraging results obtained.

OTHER CLAIMS.

Promising showings have invariably attended any measure of practical development of the mineral deposits of the Silver Creek mining district. As an evidence of this, allow me to mention some properties in addition to those described by me more at length, upon which comparatively little work has been done, yet which, nevertheless, to my mind, thoroughly merit, as fine looking claims, complete prospecting with a view toward their development into producers.

The Great Republic, owned by C. Aldrich. A vein of concentrating ore nine feet in thickness.

The Emma, owned by J. A. Ware. A vein six feet in thickness, of galena ore, with a pay streak of two feet. On this property is a tunnel forty feet long.

The Fanny, owned by Lasell & Moore. A vein ten feet in thickness, with a twenty-two-inch pay streak, assaying $25 per ton.

The Mountain Goat, owned by Scott McKrause. A vein twenty feet in thickness; assays as high as $150 in gold, silver and lead.

The Sunday Star, owned by J. N. Scott. A vein three and one-half feet thick, of iron pyrites, assaying $40 in gold and silver. On this claim Mr. Scott has sunk a shaft in ore to a depth of forty feet.

The Champion, owned by C. Cunningham. A vein twelve feet in thickness, of a concentrating ore, average asssays from which showed $20 per ton in gold, silver and lead.

The Etruria, owned by John Wesley. A vein ten feet thick, of iron pyrites and galena, assays from which showed $25 per ton of gold and silver.

The Standard, Garfield and Emma No. 2, all owned by Lloyd & Parkinson. Thickness of veins: Standard, six feet; character of ore, galena; assay value, $40 in gold, silver and lead. Garfield, six feet; character of ore, galena and iron pyrites; assay value, $30 in gold, silver and lead. Emma No. 2, six feet; character of ore, galena; assay value, $50 in gold, silver and lead.

The Murdock, owned by Murdock, Noble & Black. Thickness of vein, ten feet; character of ore, galena, quartz and iron pyrites; assay value not received.

The Cleveland, owned by Maxwell. Thickness of vein, five feet; character of ore, galena.

The Pullman, owned by Maxwell & Parkinson. Thickness of vein, three feet; character of ore, galena.

The Superior, owned by K. Anderson. Thickness of vein, ten feet; character of ore, iron pyrites.

The Annie, owned by Hubbart & Co. Thickness of vein, four feet; character of ore, gold quartz; assay value, $100 per ton in gold and silver.

STILLAGUAMISH MINING DISTRICT.

Among the many new mineral divisions born of the unprecedented yet entirely healthful interests shown this year in mining in Western Washington, is the one bearing the above title. Surely if mineral wealth lies hidden in the vast territory, part and parcel of which it is, there is wealth in Stillaguamish district awaiting the industry and practically conducted research of the many miners now at work within its confines.

Stillaguamish mining district is located in Snohomish county, in the northern part of Western Washington, and may be more particularly described by setting forth its metes and bounds as outlined in the articles of its formal organization. These are: Beginning at the mouth of Black Jack creek; thence to the southward fifteen miles; thence east to the boundaries of Silver creek and the Monte Cristo mining districts; thence north to the north fork of the Stillaguamish river; thence west fifteen miles; thence south to the place of beginning. The district was organized August 27, at a meeting held in its first camp, called Camp Independence, at the mouth of Silver gulch, at which nearly 100 miners convened. At this meeting Charles Livingston, esq., officiated as chairman, and J. W. Molique, esq., acted in the capacity of secretary. The local laws governing the new district are in conformity with those of the United States as regards mining.

Since the formal organization of this new district, research into its mineral resources has been progressing vigorously. As a result it may be said that there is every indication that the district will attain to prominence and rank with the best of the mineral divisions of Western Washington. The character of ore found is that carrying gold, silver and lead, and excellent specimens have been exhibited.

TOPOGRAPHICAL.

Described topographically, parallel spurs of mountains cross the new district from north to south; the small valleys lying between these mountains being watered by Williamson and Elk creeks, and numerous small tributaries of the Sultan river. The county, while rugged in character, is, comparatively speaking, easy of access. The visitor may take steamer from either of the two leading Sound commercial centers, proceeding to the town of Marysville; then by conveyance to Granite Falls. From the falls an excellent trail leads to Bogardus ranch, a distance of possibly ten miles. Pedestrianism from there to Hempel's ranch will next be in order; thence by footpath the traveler, following the river, reaches Camp Independence, at the mouth of Silver gulch, location of the first discoveries made in the new district. From Marysville to Silver gulch and Camp Independence the distance is not to exceed forty miles.

GEOLOGICAL.

The country rock of Stillaguamish mining district is quartz, gneiss, porphyry and highly metamorphic slates in hanging ledges, well defined, with a general trend from northeast to southwest; the general geological aspect being nearly identical with that found in the Monte Cristo mining district, previously described.

CLIMATE.

The climatic conditions of the district are identical with those of Silver Creek, Monte Cristo and the several mining districts of northwestern Washington.

WATER.

Like all the mineral divisions of northern Western Washington, the Stillaguamish has been well favored by nature in the way of water supply. Such robust streams as the Williamson and Elk creeks traverse its surface, and innumerable small brooks and springs augment an inexhaustible and superabundant supply.

TIMBER.

Timber suitable not only for mining, but for manufacturing purposes, abounds in the district. Fir, cedar, mountain pine, and the smaller species of flora thrive and are to be found throughout its area.

MINES OF THE DISTRICT.

As stated heretofore, the first camp founded in the district was Independence, at the mouth of Silver gulch, and it was in the immediate vicinage of this camp that the first locations were made. This gulch is perhaps three miles in length, and in it about twenty locations have been made since the formation of the district.

In Silver Lode gulch, a branch or tributary of the above, another great arroyo in the district, the number of locations made will number about twenty-five.

On Mineral hill, between the Sultan and Stillaguamish rivers, a large number of locations have been made; and in Boulder canyon, near this hill, some excellent looking rock has been discovered. Among the early discoveries in the district, and a ranking prospect of promise, is the Lulu claim, the property of Messrs. Smithmayer and Hemen. The Lulu is located on a vein, of which over 9,000 feet have been located.

Another of Messrs. Smithmayer and Hemen's properties, and a likely looking prospect, is on this vein. These gentlemen purpose thoroughly prospecting their discoveries, and if conditions warrant, will begin active development operations in the spring.

In Boulder canyon is located the Little Chief claim. Here a ledge of quartz ore forty feet in width has been uncovered. From samples obtained off this ledge some exceedingly encouraging returns have been received by Messrs. Williamson and Gordon, owners of the claim.

Mr. G. H. Current, of Walla Walla, who has been prospecting the district, has made several locations therein. Samples shown by this gentleman were of exceedingly fine looking ore.

SNOQUALMIE MINING DISTRICT.

Situated in the far eastern portion of King county is this comparatively new mining district. Its area comprises all of that territory on the Cascade mountains drained by the south, middle and north forks of the Snoqualmie river, and extending from the foothills of its western flank to the summit of that range. This district owes its existence to Messrs. W. C. Weeks, Herbert Powers, Sam. D. Gustin, D. H. Rushing, J. W. Guye, George W. Tibbetts, Fred B. Weistling, Charles F. Blackburn and others, being formally organized by these gentlemen early in the spring of 1891.

Geologically, the Snoqualmie mining district is found to present the same general aspect as the country contiguous thereto, which I have heretofore described at length. This district is about thirty miles square, containing in the aggregate 900 square miles.

Gold bearing quartz comprises the principal minéral bearing rock, and some promising looking prospects are to be found in the district. But a small portion of the district has ever been explored, to say nothing of prospecting. Locations made in the prospected area comprise about 150 in number, all of which seem to merit continued development.

The principal seat of present operations in the district seems to be about the claims owned by the Cascade Gold Mining and Milling Company, a corporation which seems to have gone to work with a will to ascertain the true mineral value of the area. This company is developing three claims. The ore extracted is gold and silver, and returns as high as $1,000 per ton are said to have been the result of assays of samples of it. Several large bodies of ore have been exposed, ranging in width from six to sixty feet. An excellent tunnel site is at hand, and the claims are being vigorously developed by the manager of the company, Mr. Charles F. Blackburn.

One of this company's claims, called the Copper Chief, a galena-carbonate and copper producer, near the Snoqualmie pass, shows a large body of chalcopyrite and lead-silver ore. The first-class copper ore is said to have assayed $103 per ton in gold, silver and copper, the best of the galena-carbonate ore returning as high as $374 per ton in gold, silver and lead. Assays of other galena-silver veins have returned $50 to $130 per ton.

R. L. Blackburn is the discoverer of three very promising prospects in the district, known as the Galena Chief, Seven-Twenty and Ingersoll claims. On each an excellent showing in the way of surface outrop of ore is to be seen. Samples of this ore (quartz and galena) have returned from $18 to $140 per ton.

Paying placers are also to be found in this district, along the middle fork of the Snoqualmie river, where three claims were recently located by Messrs. Charles Blackburn, Ben Finnel and J. J. Watts.

On Uncle Si mountain, one and one-half miles from the town of North Bend, across the middle fork of the Snoqualmie river, in the district, is located the Black Jack mine, the claim of some North Bend and Seattle operators. Of this property Charles F. Blackburn is also the general manager. Several tunnels have been run in for short distances on the vein, good showings being the result. The claim next north of the above is the Galena Star, where a tunnel has been driven 200 feet for the purpose of tapping the vein.

On the south fork of the Snoqualmie river, where is located the property of the West Gate Gold and Silver Mining Co., with the veteran miner, J. W. Miller, in charge as general manager. Mr. Miller reports a very encouraging outlook as regards these properties.

The Seattle Mining and Development Co., Messrs. Sherk, Baxter, Blackburn and White, and the Snoqualmie Pass Gold, Silver and Marble Mining and Milling Co. own good looking prospects in this vicinity, and all have been hard at work the past autumn developing them.

MINES AND MINERALS OF WASHINGTON.

PART SIXTH.

SECOND ANNUAL REPORT OF GEORGE A. BETHUNE,
STATE GEOLOGIST.

WASHINGTON PLACERS.

The date of the discovery of gold in Washington—in fact throughout the mineral bearing area of the Pacific Northwest—dates back many years, and is identical, I might say here, with the discovery of paying deposits of gravel made by California gold hunters of the early days along the various streams in the western-northern, eastern-northern and central portions of the then styled wilderness of Washington Territory.

As far as can be with accuracy ascertained, gold was first discovered in Western Washington away up near the headwaters of Skagit river, and near the British Columbia boundary line. The date of this discovery may be chronicled as the spring of 1868, and as nearly as I can ascertain, Thomas Keefe, now a fire commissioner in one of the California cities of prominence; William Milliken, a veteran gold seeker, and Joshua Hardy, are responsible for this discovery. In the early days the attention of the miners in that region was entirely engrossed with the work of delving gold from placer claims. Naturally, in this work, as is the case to-day, the most simple and inexpensive arts known to mining were employed.

The newcomer to the Washington placers might be denoted, to use the old expression, "by the tools he used." These generally comprised the "rocker" of ye olden and modern time as well; a tin can fastened to a wooden handle, to serve the purpose of a dipper; a gold pan, crevassing pick, shovel and sometimes a short pinch bar. As his wealth increased, so, as a general thing, did the appliances used in the pursuance of his fascinating vocation, with the placer miner. The rocker, did the gravel pan out rich, was superseded in time by the "long tom," the latter eventually becoming the predecessor of the sluice box and its incidental belongings.

There is no gainsaying the fact that the wealth of the early Skagit placers was enormous. Old timers are fond of relating the stories of rich rewards resultant on a comparatively small measure of manual labor in the placer fields of the Skagit. As closely as I can calculate from accurate accounts that have come into my possession from reliable sources, the Skagit placers have yielded as high, on an average, as $28 per cubic yard; and instances have been cited me wherein miners have secured three and even five times that amount, but from what I can glean such instances were of rare occurrence.

While the Skagit placers may now be said to be well nigh depleted of their valuable contents, it is the fact that to this day on portions of the stream placers are being worked with good results by many miners. On Ruby creek, a tributary of the Skagit river, some fabulously wealthy placers were found, and here, too, placer mining is now being carried on with profitable results.

Possibly the placer deposits in Western Washington now engrossing most attention are those located along the Sultan, Snoquamie, Stillaguamish and Raging rivers, in northern Western Washington. Estimates place the yield per cubic yard of Sultan river placers, taken generally the past pear, at from forty to eighty-five cents. The Sultan placers have been represented this year by about 150 legally located claims, the district extending the whole length almost of both sides of that stream from its source near Silver creek to its confluence with the Skykomish river, a distance of about eighteen miles from the city of Snohomish, in Snohomish county.

Like those of the Skagit, the placers of the Sultan were discovered years ago, the first gold being taken from gravel on the latter stream as far back as 1869. Like the Skagit deposits, the larger portion of the paying ground along the Sultan river may be said to be of glacier formation; and undoubtedly the natural deepening of their bed, an incident throwing the deposits to their sides, has had much to do with the deposition of gold from the bed-rock along their shores.

Aside from the placer deposits of the Skagit and Sultan rivers are those to be found along the Stillaguamish, Snoqualmie and Raging rivers. On the former are some very rich deposits, and the work done in the fields of the Snoqualmie has been attended with an amount of profit sufficient to warrant miners there in embarking in the undertaking next year on a more extensive scale than ever.

On the eastern side of the Cascade mountains are fourteen different streams, along the banks and in the immediate vicinage of which paying placers have been worked for years. I refer to the

Cle-Elum river, Teanaway river and Swauk river, streams tributary to the Yakima; to the upper waters of that river itself, and to the Icicle and Peschastin rivers, which empty into the Wenatchee river; to the Entiat river, emptying into the Columbia, and the Stehekin, emptying into Lake Chelan; and to the Methow, Okanogan, Nespilem, San Puel, Spokane, Colville and Kettle rivers, all directly or indirectly tributary to the great main waterway, the Columbia.

Placer gold was found in Eastern Washington long antecedent to its discovery in the western half of the state. In fact, gold was found along the streams in the middle, northern and northeastern portions of this state way back in the early fifties, and it is believed the aborigines knew of its existence there at even an earlier period. The placers of O'Sullivan creek and Similkameen river have been known these five and thirty years or more to white men. From these placers hundreds of thousands of dollars have been taken, and Chinese miners are said to be working them with profit to this day.

The Cle-Elum river placers have been profitable fields of labor for years, and this year has witnessed the best showing made in a decade in these workings. Many white miners are operating there, and as far as I can ascertain, are meeting with more than a fair measure of success in their efforts. The Cle-Elum placers may be said to extend from near the headwaters of that stream, in the vicinity of Snoqualmie pass, to a point half its length in the direction of its confluence with the Yakima river.

The Teanaway is a comparatively small stream, yet richly endowed in its shore line with paying gravel. More attention than ever has been this year given placer mining on the Teanaway, and, I am glad to state, results accruing from work done have been of a very flattering character.

The Swauk placers are famous throughout the country. I take it that more profit has been realized from placer mining along this fine stream in recent years than in all the other deposits of placer ground in the state combined. The past year has witnessed a decided boom in this species of mining on the Swauk, and from early spring until late the past autumn hundreds of miners have found lucrative reward for their work along the Swauk.

The Icicle and Peschastin rivers have also long been known as rich placer fields, the discovery of placer gold on the latter stream being made at a very early period in the history of the state as a territory. These fields were also operated the past year with great success, notably the Icicle, where a great quantity of gold has been secured, amongst it many nuggets of great value.

The Entiat and Stehekin placers may be chronicled comparatively recent discoveries. Both deposits are, however, rich in mineral. On the Stehekin river about 120 miners have found lucrative employment the past season, and about half that number report an excellent season's work on the Entiat river. The Methow, Okanogan, Nespilem, San Puel, Spokane, Colville and Kettle river placers are old discoveries, and yet far from being worked out. Chinese mainly derive a revenue from these last named deposits. While it is impossible to obtain from these people anything approximating an accurate idea of the monitary results of their labors, the fact that they steadily continue to work, and are with an energy and enterprise commendable continually improving methods for obtaining the gold, speaks volumes in favor of the presumption that these old placers are still rich in the precious metal.

On the Columbia river, from the Little Dalles, near the British Columbia boundary line, to Pasco, in Franklin county, the placers known to be valuable thirty years ago are being worked with profit to this day, but, in the majority of instances, by Chinese miners. Valuable placers are also to be found along the Sultan river.

I may say the placer fields of this state are yet among the most important of the factors destined to upbuild it. Old as some of these are, well worked as the majority of them are, they still offer a most inviting field to the miner with whom this character of gold digging, taking into consideration its comparative inexpensiveness, must ever be popular.

NEW MINERAL DISTRICTS.

I have to report the formation of two new mineral districts, each of much promise, formed during the year just passed.

Gold Hill mining district is the title given by the founders to a considerable area of mineral land situated in the mountains of Yakima county. So late in the season were discoveries made warranting the formation of this district, that aside from cursory prospecting by the founders, but little actual development work was accorded the territory. There is every prospect that with the opening of the season this year, Gold Hill district will be given the attention which present prospects certainly indicate should be awarded it.

The second new field of operations has been named by its founders Mineral Creek mining district. This new district is located on Mineral creek, a tributary of the Nesqually river, in Lewis county, Western Washington. It has already attracted a deal of attention from ambitious novitiates in mining on this side of the Cascade range. I have seen excellent samples of galena ore, rich in silver, and carrying a heavy percentage of lead, shown me by Messrs. Evans and Davis, discoverers of the new camp. Judging from present indications, it is likely this district will be thoroughly prospected, and if circumstances warrant, be considerably developed during the present year.

NEW MINERALS.

I have to report the discovery in this state of a deposit of mineral of value, hitherto believed not to exist in Washington. This office returns its thanks to the discoverers for the commendable promptness exercised in acquainting the state geologist of their discovery. I shall here briefly refer to it.

CHROME ORE.

As is well known, chrome ore is esteemed of great value in certain lines of manufacture, especially that of chrome steel and particolored paints. The intrinsic value of the ore is always enhanced when the proportion of chromic oxide contained is found to be large.

Two discoveries of deposits of chrome ore, which certainly warrant careful practical development, have been made in Skagit and Kittitas counties. I have seen samples of ore from these deposits, and have no hesitancy in recomending that their owners give them a thorough development. I append an analysis of a sample of this ore:

ANALYSIS.

	Per cent.
Chromic oxide Cr. O. (sesqui-oxide of chromium)	51.28
Protoxide of iron	12.55
Manganese oxide	1.16
Alumina oxide	20.12
Lime	5.19
Magnesia	.30
Silica	9.40
Total	100.00
Per cent. of chromium	35.13

Specimens of asbestos, graphite, nickel, molybdenum and cinnabar have been received at this office, but the manifold duties incumbent upon me have precluded an extended investigation as regards these minerals.

MINES AND MINERALS OF WASHINGTON.

PART SEVENTH.

SECOND ANNUAL REPORT OF GEORGE A. BETHUNE, STATE GEOLOGIST.

ANALYTICAL DEMONSTRATION.

COMPUTATIVE RESULTS—WEIGHT AND VOLUME OF MINERALS.

In my last annual report (Mines and Minerals of Washington, 1890) I treated of the method of analyzation of gold and silver bearing ores in so far as the method of ascertaining the value in gold and silver of these was concerned. I also published tables of computative results, and weight and volume of a number of the more important minerals.

I have been requested to herein republish these matters, and cheerfully comply.

About ten ounces of the ore should be taken and crushed to a moderate degree of fineness in an iron mortar. It is then passed through a sieve until it assumes the same degree of fineness throughout. The powder is then placed on a sheet of paper, usually about 10 x 12 inches. A part of the powder is then taken—generally about one-fourth of an ounce, troy weight—and put into a scorifier or a small fire clay cup that is half filled with granulated lead. The ore and lead are then covered with borax. The scorifier is put into a muffle or fire clay oven, heated to a white heat for twenty or thirty minutes. This operation is only complete when the ore and slag have become liquid. The mass is then poured into an iron mould and allowed to cool. When cold, a button of metal will be found with slag adhering to it. With a hammer, remove the slag and pound the button into a cube. Now take a cupel or a little cup made of bone ash, and place in the muffle, being careful to keep a high heat; place the button in the cupel. The button will melt if the cupel is hot, and the lead will be driven off, leaving another button composed of gold and silver. This operation generally takes about thirty minutes. Pick up the button when cool and weigh it.

It is generally desired to know at what ratio to each other the gold and silver exist, or what part of the button is of gold and what part of silver. The button is now dropped into a test tube, which is a small glass tube, closed at one end, and ten drops of nitric acid dropped in. Heat the tube slowly over a spirit lamp. When the silver has dissolved, fill the tube with water, invert it over a small cup of fire clay, made to fit the top of the test tube, called annealing cup, and in a few minutes the gold will settle at the bottom. Pour off the silver solution and dry the cup containing the gold over the spirit lamp. Weigh the button and find what ratio of weight it bears to the whole button.

If the button of gold and silver originally weighed twenty grains, and the gold remaining in the test tube after the last operation, called refining, weighs two grains, the silver must have weighed eighteen grains, or gold exists in the ratio of two to twenty, or one-tenth, and silver eighteen to twenty, or nine-tenths; and if as shown heretofore, in a ton of ore there are 233.33 ounces of gold and silver, there must be 23.33 ounces of gold and 210 ounces of silver.

The Troy is the standard system of weights in use in cases of this character. Appended will be found a table showing the amount of gold and silver, in ounces and decimals, contained in one ton of ore of 2,000 pounds, from the weight of this same gold and silver, obtained in an assay of (half an ounce) 240 grains of ore. My principle of construction of this compilation is to take the thousandth part of ten grains, which is, of course, equal to the hundredth part of one grain. If 240 grains of ore yield .01 grains, then 240 pounds of the same ore will yield .01 pounds, and 2,000 pounds will yield 8.333 times as much, for $\frac{2000}{240}=8.333$. One pound avoirdupois contains 14.58333 troy ounces. One hundredth of this, .145833, which, multiplied by 8.333, equals 1.2152729 troy ounces, which is the starting point of the table.

COMPUTATIVE RESULTS.

If two hundred and forty grains of ore give a gold and silver button weighing thousanths of a unit of ten grains.	*One ton of ore, gold and silver, will yield in ounces......*	*Value of silver per ton, at $1.29 per ounce......*	*Value of gold per ton, at $20.67 per ounce......*
.001	1.21	$1 56	$25 01
.002	2.43	3 14	50 23
.003	3.64	4 71	75 24
.004	4.86	6 28	100 46
.005	6.08	7 86	125 68
.006	7.29	9 42	150 70
.007	8.51	10 99	175 92
.008	9.72	12 57	200 93
.009	10.94	14 14	226 15
.010	12.15	15 69	251 16
.011	13.37	17 29	276 38
.012	14.58	18 85	301 39
.013	15.80	20 43	326 61
.014	17.01	21 99	351 63
.015	18.23	23 57	376 85

EXPLANATORY.—Say, for instance, .010 of gold has been found. Note that decimal in the left-hand column of the table. One ton of ore in ratio to .010 will yield the number of ounces noted in the next column to the right, directly opposite .010, and so on through the table. Computation with .010 taken as a basis may easily be made as high as circumstances may require.

WEIGHT AND VOLUME OF MINERALS.

Appended is a table showing the weight of one cubic foot and the volume of one ton of a number of the more important minerals. One cubic foot of water weighs sixty-two pounds:

Minerals.	*Weight of cubic foot in pounds*	*Cubic feet in one ton, 2,000 lbs.*
Quartz	162	12.34
Silver glance	455	4.39
Ruby silver	362	5.52
Brittle silver	386	5.18
Horn silver	345	5.80
Antimony glance	287	6.99
Cinnabar	549	3.64
Copper pyrites	262	7.63
Gray copper	280	7.14
Galena	461	4.34
Zinc blende	249	8.03
Iron pyrites	312	6.41
Limestone	174	11.50
Clay	162	12.34

FOR COURTESIES EXTENDED

Me in my official capacity as State Geologist, I beg leave to return thanks to the following named enterprising and progressive citizens of this and adjoining states:

Hon. W. H. IRELAN, jr., State Mineralogist of California.

JONATHAN BOURNE, jr., Portland, Or.

J. M. BURKE, Burke, esq., Idaho.

C. H. PRESCOTT, Vice President Northern Pacific Railroad Co., Tacoma, Wash.

JOHN L. HOPKINS, esq., Agent Northern Pacific Express Co., Tacoma, Wash.

Col. JAMES F. WARDNER, Fairhaven, Wash.

Hon. J. W. SLATER, Colville, Wash.

Hon. HENRY DRUM, Tacoma, Wash.

Col. THOMAS EWING, Seattle, Wash.

L. W. GETCHELL, esq., Seattle, Wash.

F. W. DUNN, esq., Seattle, Wash.

A. P. WEBB, esq., Seattle, Wash.

WESLEY WILSON, esq., Seattle, Wash.

ANGUS McINTOSH, esq., Seattle, Wash.

NELSON BENNETT, esq., Tacoma, Wash.

ROBERT WINGATE, esq., Tacoma, Wash.

W. R. RUST, esq., Manager Tacoma Smelter.

ALLEN C. MASON, esq., Tacoma, Wash.

J. T. McDONALD, esq., Ellensburgh, Wash.

A. C. COWHERD, esq., Loomiston, Wash.

THOMAS ISMAY, esq., Superintendent Bucoda coal mine.

PATRICK CLARK, esq., Superintendent Poorman mine, Idaho.

SAML. COULTER, esq., Portland, Or.

DONALD FERGUSON, esq., Seattle, Wash.

GEO. E. PFUNDER, Superintendent Washington's Mineral Exhibit at World's Columbian Exposition.

I desire, too, in the name of our mining community, to return thanks to the press throughout the state for the sincere and kindly interest manifestd in the upbuilding and developing of the mineral interests of our commonwealth.

Respectfully submitted.

GEORGE A. BETHUNE,
State Geologist.

INDEX.

Lightning Source UK Ltd.
Milton Keynes UK
UKHW010609120219
337137UK00007B/1464/P